Ben Stacy Jerrik (Ed.)

Rokitnik, Lidzbark County

Ben Stacy Jerrik (Ed.)

Rokitnik, Lidzbark County

Kiwity, Warmian-Masurian Voivodeship

Part Press

Imprint

Permission is granted to copy, distribute and/or modify this document under the terms of the GNU Free Documentation License, Version 1.2 or any later version published by the Free Software Foundation; with no Invariant Sections, with the Front-Cover Texts, and with the Back- Cover Texts. A copy of the license is included in the section entitled "GNU Free Documentation License".

All parts of this book are extracted from Wikipedia, the free encyclopedia (www.wikipedia.org).

You can get detailed informations about the authors of this collection of articles at the end of this book. The editors (Ed.) of this book are no authors. They have not modified or extended the original texts.

Pictures published in this book can be under different licences than the GNU Free Documentation License. You can get detailed informations about the authors and licences of pictures at the end of this book.

The content of this book was generated collaboratively by volunteers. Please be advised that nothing found here has necessarily been reviewed by people with the expertise required to provide you with complete, accurate or reliable information. Some information in this book maybe misleading or wrong. The Publisher does not guarantee the validity of the information found here. If you need specific advice (f.e. in fields of medical, legal, financial, or risk management questions) please contact a professional who is licensed or knowledgeable in that area.

Any brand names and product names mentioned in this book are subject to trademark, brand or patent protection and are trademarks or registered trademarks of their respective holders. The use of brand names, product names, common names, trade names, product descriptions etc. even without a particular marking in this works is in no way to be construed to mean that such names may be regarded as unrestricted in respect of trademark and brand protection legislation and could thus be used by anyone.

Cover image: www.ingimage.com
Concerning the licence of the cover image please contact ingimage.

Publisher:
Part Press is a trademark of
International Book Market Service Ltd., 17 Rue Meldrum, Beau Bassin, 1713-01 Mauritius
Email: info@bookmarketservice.com
Website: www.bookmarketservice.com

Published in 2012

Printed in: U.S.A., U.K., Germany. This book was not produced in Mauritius.

ISBN: 978-613-8-66851-0

Contents

Rokitnik, Lidzbark County

Rokitnik

— Village —

Rokitnik

Coordinates: 54°7′N 20°47′E

Country	▬ Poland
Voivodeship	Warmian-Masurian
County	Lidzbark
Gmina	Kiwity

Rokitnik [rɔˈkitnik] is a village in the administrative district of Gmina Kiwity, within Lidzbark County, Warmian-Masurian Voivodeship, in northern Poland.[1] It lies approximately 3 kilometres (**unknown operator: u'strong'** mi) north-east of Kiwity, 14 km (**unknown operator: u'strong'** mi) east of Lidzbark Warmiński, and 42 km (**unknown operator: u'strong'** mi) north-east of the regional capital Olsztyn.

Before 1772 the area was part of Kingdom of Poland, 1772-1945 Prussia and Germany (East Prussia).

References

[1] "Central Statistical Office (GUS) - TERYT (National Register of Territorial Land Apportionment Journal)" (http://www.stat.gov.pl/broker/access/prefile/listPreFiles.jspa) (in Polish). 2008-06-01. .

East Prussia

East Prussia (German: *Ostpreußen*, pronounced ['ɔst‿pʀɔYsən] (🔊 listen); Polish: *Prusy Wschodnie*; Lithuanian: *Rytų Prūsija* or *Rytprūsiai*; Russian: Восточная Пруссия or *Vostochnaya Prussiya*; Latin: *Borussia orientalis*) is the main part of the region of Prussia along the southeastern Baltic Coast from the 13th century to the end of World War II in May 1945.[1] From 1772–1829 and 1878–1945, the Province of East Prussia was part of the German state of Prussia. The capital city was Königsberg.

East Prussia enclosed the bulk of the ancestral lands of the Baltic Old Prussians. During the 13th century, the native Prussians were conquered by the crusading Teutonic Knights. The indigenous Balts who survived the conquest were gradually converted to Christianity. Because of Germanization and colonisation over the following centuries, Germans became the dominant ethnic group, while Poles and Lithuanians formed minorities. From the 13th century, East Prussia was part of the monastic state of the Teutonic Knights, which became the Duchy of Prussia in 1525.[2] The Old Prussian language had become extinct by the 17th or early 18th century.[3]

Following the death of Hohenzollern Albert of Brandenburg Prussia, Duke of Prussia (1525–1568), Joachim II, the prince-elector Kurfürst of Brandenburg, became co-inheritor of Ducal Prussia. In 1577, House of Hohenzollern co-regents took over administration from Albert's only son, Albert Friedrich. In 1618 the Duchy of Prussia again passed by inheritance and in personal union with the Hohenzollerns of Brandenburg and the territory was called Brandenburg-Prussia. The territories of the House of Hohenzollern were scattered in Franconia, Brandenburg, eastern Prussia and elsewhere.

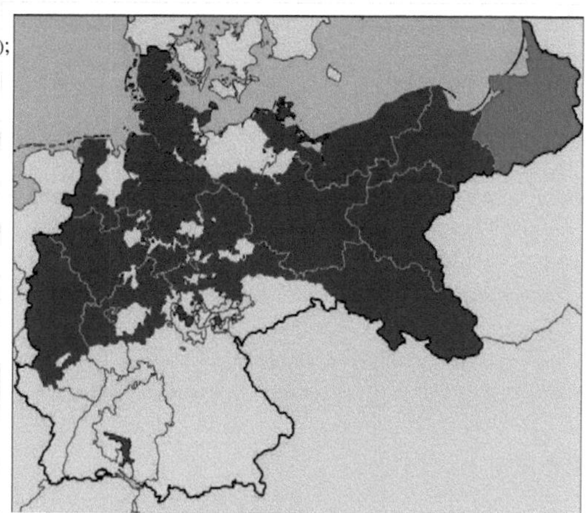

The Province of East Prussia (red), within the Kingdom of Prussia, within the German Empire, as of 1871.

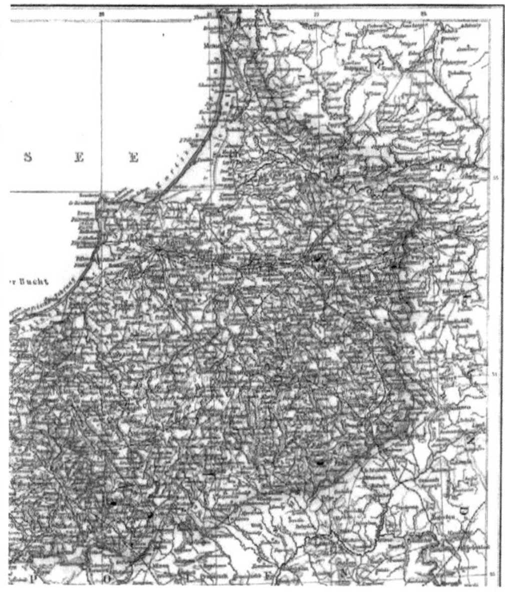

Map of East Prussia in 1881.

Because the duchy was outside of the core Holy Roman Empire (Prussia was under HRE administration by the Teutonic Order grandmasters), the prince-electors of Brandenburg were able to proclaim themselves kings *in* Prussia beginning in 1701. After the annexation of most of western Royal Prussia in the 1772 First Partition of the Polish-Lithuanian Commonwealth, East Prussia was connected by land with the rest of the Prussian state and was reorganized as the Province of East Prussia the following year. Between 1829 and 1878, the Province of East Prussia was joined with West Prussia to form the Province of Prussia.

The Kingdom of Prussia became the leading state of the German Empire after its creation in 1871. However, the Treaty of Versailles following World War I restored West Prussia to Poland and made East Prussia an exclave of Weimar Germany, while the Memel Territory was detached and was annexed by Lithuania in 1923. Following Nazi Germany's defeat in World War II in 1945, war-torn East Prussia was divided at Joseph Stalin's insistence between the Soviet Union (the Kaliningrad Oblast in the Russian SFSR and the constituent counties of the Klaipėda Region in the Lithuanian SSR) and the People's Republic of Poland (the Warmian-Masurian Voivodeship).[4] The capital city Königsberg was renamed Kaliningrad in 1946. The German population of the province was largely evacuated during the war or expelled shortly thereafter in the expulsion of Germans after World War II. An estimated 300,000 (around one fifth of the population) died either in war time bombings raids or the battles to defend the province.

History

From Catholic monastic state to Protestant duchy

Upon the invitation of Duke Konrad I of Masovia, the Teutonic Knights took possession of Prussia in the 13th century and created a monastic state to administer the conquered Old Prussians. The Knights' expansionist policies brought them into conflict with the Kingdom of Poland and embroiled them in several wars, culminating in the Polish-Lithuanian-Teutonic War, whereby the united armies of Poland and Lithuania, defeated the Teutonic Order at the Battle of Grunwald (Tannenberg) in 1410. Its defeat was formalised in the Second Treaty of Thorn in 1466 ending the Thirteen Years' War, and leaving the former Polish region Pomerelia and under Polish control. Together with Warmia it formed the province of Royal Prussia. Eastern Prussia remained under the Knights, but as a fief of Poland. 1466 and 1525 arrangements by kings of Poland were not verified

The fortress *Ordensburg Marienburg*, founded in 1274, the world's largest brick castle and the Teutonic Order's headquarters on the River Nogat.

by the Holy Roman Empire as well as the previous gains of the Teutonic Knights were not verified.

The Teutonic Order lost eastern Prussia when Grand Master Albert of Brandenburg-Ansbach converted to Lutheranism and secularized the Prussian branch of the Teutonic Order in 1525. Albert established himself as the first duke of the Duchy of Prussia and a vassal of the Polish crown by the Prussian Homage. Walter von Cronberg, the next Grand Master, was enfeoffed with the title to Prussia after the Diet of Augsburg in 1530, but the Order never regained possession of the territory. In 1569 the Hohenzollern prince-electors of the Margraviate of Brandenburg became co-regents with Albert's son, the feeble-minded Albert Frederick.

The Administrator of Prussia, the grandmaster of the Teutonic Order Maximilian III, son of emperor Maximilian II died in 1618. Albert's line died out in 1618, and the Duchy of Prussia passed to the Electors of Brandenburg, forming Brandenburg-Prussia. Through the treaties of Wehlau, Labiau, and Oliva, Elector and Duke Frederick William succeeded in revoking king of Poland's sovereignty over the Duchy of Prussia in 1660. The absolutist elector also subdued the noble estates of Prussia.

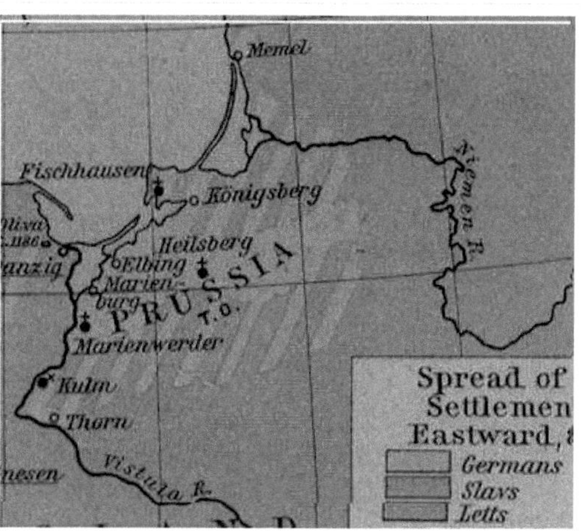

Ethnic settlement in East Prussia by the 14th century.

Monument of Grand Master Albert, the first Duke of Prussia; Malbork, Poland

Kingdom of Prussia

Although Brandenburg was a part of the Holy Roman Empire, the Prussian lands were not within the Holy Roman Empire and were with the administration by the Teutonic Order grandmasters under jurisdiction of the Emperor. In return for supporting Emperor Leopold I in the War of the Spanish Succession, Elector Frederick III was allowed to crown himself "King in Prussia" in 1701. The new kingdom ruled by the Hohenzollern dynasty became known as the Kingdom of Prussia. The designation "Kingdom of Prussia" was gradually applied to the various lands of Brandenburg-Prussia. To differentiate from the larger entity, the former Duchy of Prussia became known as *Altpreußen* ("Old Prussia"), the province of Prussia, or "East Prussia".

Approximately one-third of East Prussia's population died in the plague and famine of 1709–1711,[5] including the last speakers of Old Prussian.[6] The plague, probably brought by foreign troops during the Great Northern War, killed 250,000 East Prussians, especially in the province's eastern regions. Crown Prince Frederick William I led the rebuilding of East Prussia, founding numerous towns. Thousands of Protestants expelled from the Archbishopric of Salzburg were allowed to settle in depleted East Prussia. The province was overrun by Imperial Russian troops during the Seven Years' War.

After the First Partition of Poland in 1772, Warmia, part of western Royal Prussia, was merged with the former Duchy of Prussia. On 31 January 1773, King Frederick II announced that the newly annexed lands were to be known as the Province of West Prussia, while the former Duchy of Prussia and Warmia became the Province of East Prussia.

From 1824–1878, East Prussia was combined with West Prussia to form the Province of Prussia, after which they were reestablished as separate provinces.

German Empire

Along with the rest of the Kingdom of Prussia, East Prussia became part of the German Empire during the unification of Germany in 1871.

From 1885 to 1890 Berlin's population grew by 20%, Brandenburg and the Rhineland gained 8.5%, Westphalia 10%, while East Prussia lost 0.07% and West Prussia 0.86%. This stagnancy in population despite a high birth surplus in eastern Germany was because many people from the East Prussian countryside moved westward to seek work in the expanding industrial centres of the Ruhr Area and Berlin (see *Ostflucht*).

The population of the province in 1900 was 1,996,626 people, with a religious make up of 1,698,465 Protestants, 269,196 Roman Catholics, and 13,877 Jews. The Low Prussian dialect predominated in East Prussia, although High Prussian was spoken in Warmia. The numbers of Masurians and Prussian Lithuanians decreased over time due to the process of Germanization. The Polish-speaking population concentrated in the south of the province (Masuria and Warmia) and all German geographic atlases at the start of 20th century showed the southern part of East Prussia as Polish with the number of Poles estimated at the time to be 300.000.[7] Lithuanian-speaking Prussians concentrated in the northeast (Lithuania Minor). The Old Prussian ethnic group became completely Germanized over time and the Old Prussian language died out in the 18th century.

World War I

At the beginning of World War I, East Prussia became a theatre of war when the Russian Empire invaded the country. The Russian Army encountered at first little resistance because the bulk of the German Army had been directed towards the Western Front according to the Schlieffen Plan. In the Battle of Tannenberg in 1914 and the Second Battle of the Masurian Lakes in 1915, however, the Russians were decisively defeated and forced to retreat, followed by the German Army advancing into Russian territory. The majority of the civilian population fled before the invading Russian Army, while several thousand remaining civilians were deported to Russia. Treatment of civilians by both armies was mostly disciplined, although 74 civilians were killed by Russian troops in the Abschwangen massacre. The region had to be rebuilt because of damage caused by the war.

Weimar Republic

With the forced abdication of Emperor William II in 1918, Germany became a republic. Most of West Prussia and the former Prussian Province of Posen, territories annexed by Prussia in the 18th century Partitions of Poland, were ceded to the Second Polish Republic according to the Treaty of Versailles. East Prussia became an exclave, being separated from mainland Germany. The Seedienst Ostpreußen was established to provide an independent transport service to East Prussia.

On 11 July 1920, amidst the backdrop of the Polish-Soviet War, the East Prussian plebiscite in eastern West Prussia and southern East Prussia was held under Allied supervision to determine if the areas should join the Second Polish Republic or remain in Weimar Germany Province of East Prussia. 96.7% of the people voted to remain within Germany (97.89% in the East Prussian plebiscite district).

East Prussia from 1923 to 1939 between the wars

The Klaipėda Territory, a League of Nations mandate since 1920, was occupied by Lithuanian troops in 1923 and was annexed without giving the inhabitants a choice by the ballot.

Nazi Germany

In 1932 the local paramilitary SA had already started to terrorise their political opponents. On the night of 31 July 1932 there was a bomb attack on the headquarters of the Social Democrats in Königsberg, the Otto-Braun-House. The Communist politician Gustav Sauf was killed, the executive editor of the Social Democrat *"Königsberger Volkszeitung"*, Otto Wyrgatsch, and the German People's Party politician Max von Bahrfeldt were severely injured. Members of the Reichsbanner were attacked and the local Reichsbanner Chairman of Lötzen, Kurt Kotzan, was murdered on 6 August 1932.[8] [9] After the Nazis took power in Germany, opposition politicians were persecuted and newspapers were banned. The Otto-Braun-House was requisitioned and became the headquarter of the SA that used the house to imprison and torture opponents. Walter Schütz, a communist Member of the Reichstag was murdered here.[10]

In 1938 the Nazis altered about one-third of the toponyms of the area, eliminating, Germanizing, or simplifying a number of Old Prussian names, as well as those Polish or Lithuanian names originating from refugees to Prussia during and after the Protestant Reformation. More than 1,500 places were ordered to be renamed by 16 July 1938 following a decree issued by gauleiter and Oberpräsident Erich Koch and initiated by Adolf Hitler.[11] Many who would not co-operate with the rulers of Nazi Germany were sent to concentration camps and held there prisoner until their death or liberation.

World War II

Partitions of Eastern Europe before, during, and after World War II (Map is written in German)

In 1939 East Prussia had 2.49 million inhabitants, 85% of them ethnic Germans, the others Poles in the south who, according to Polish estimates numbered in the interwar period around 300,000-350,000,[12] and Lietuvininkai speaking Lithuanian (Baltic) in the northeast. Most German East Prussians, Masurians, and Lietuvininkai were Lutheran, while the population of Ermland was mainly Roman Catholic due to the history of its bishopric. The East Prussian Jewish Congregation declined from about 9,000 in 1933 to 3,000 in 1939, as most fled from Nazi rule.[13] Those who remained were later deported and killed in the Holocaust.

In 1939 the Regierungsbezirk Zichenau was annexed by Germany and incorporated into East Prussia. Despite Nazi propaganda presenting all of the regions annexed as possessing significant German populations that wanted reunification with Germany, the Reich's statistics of late 1939 show that only 31,000 out of 994,092 people in the annexed Polish western territories were ethnic Germans.

East Prussia was only slightly affected by the war until January 1945, when it was devastated during the East Prussian Offensive. Most of its inhabitants became refugees in bitterly cold weather during the Evacuation of East Prussia.

Evacuation of East Prussia

In 1944 the medieval city of Königsberg, which had never been severely damaged by warfare in its 700 years of existence, was almost completely destroyed by two Allied air raids on the night of 26/27 August 1944 and three nights later on the 29/30 August 1944. Winston Churchill (*The Second World War*, Book XII) had erroneously believed it to be "a modernised heavily defended fortress" and ordered its destruction.

Gauleiter Erich Koch protracted the evacuation of the German civilian population until the Eastern Front approached the East Prussian border in 1944. The population had been systematically misinformed by *Endsieg* Nazi propaganda about the real military state of affairs. As a result many civilians fleeing westward were overtaken by retreating Wehrmacht units and the rapidly advancing Red Army.

Reports of Soviet atrocities in the Nemmersdorf massacre of October 1944 and organised rape spread fear and desperation among the civilians. Thousands lost their lives during the sinkings by Soviet submarine of the refugee ships *Wilhelm Gustloff*, the *Goya*, and the *General von Steuben*. Königsberg surrendered on 9 April 1945, following the desperate four-day Battle of Königsberg. The number of civilians killed is estimated to be at least 300,000 with most dying under horrible conditions.

However, most of the German inhabitants, which then consisted primarily of women, children, and old men, did manage to escape the Red Army as part of the largest exodus of people in human history.[14] "A population which had stood at 2.2 million in 1940 was reduced to 193,000 at the end of May 1945."[15]

Expulsion of Germans from East Prussia after World War II

Shortly after the end of the war in May 1945, Germans who had fled in early 1945 tried to return to their homes in East Prussia. An estimated number of 800,000 Germans were living in East Prussia during the summer of 1945.[16] Many more were prevented from returning, and the German population of East Prussia was almost completely expelled by the communist regimes. During the war and for some time thereafter forty-five camps for about 200 to 250 thousand forced labourers were established, the vast majority of them were deported to the Soviet Union, including the Gulag camp system.[17] The largest camp with about 48,000 inmates was established at Deutsch Eylau (Ilawa).[17]

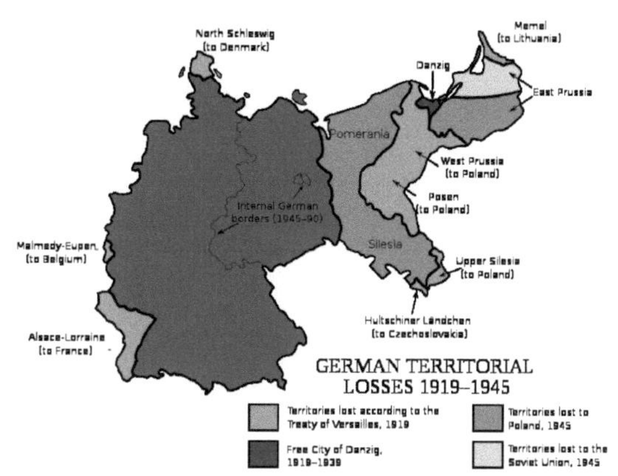

Germany's shift of borders after the world wars, dividing East Prussia among other countries.

Southern part to Poland

Representatives of the Polish government officially took over the civilian administration of the southern part of East Prussia on 23 May 1945.[17] Subsequently Polish expatriates from Polish lands annexed by the Soviet Union as well as Ukrainians from southern Poland, expelled throughout Operation Vistula in 1947, were settled in the southern part of East Prussia, now the Polish Warmian-Masurian Voivodeship. In 1950 the Olsztyn Voivodeship counted 689,000 inhabitants, 22.6% of them coming from areas annexed by the Soviet Union, 10% Ukrainians, and 18.5% of them pre-war inhabitants. The remaining pre-war population was treated as Germanized Poles and a policy of re-Polonization was pursued throughout the country[18] Most of these "Autochthones" chose to emigrate to West Germany from the 1950s through 1970s (between 1970 and 1988 55,227 persons from Warmia and Masuria moved to Western Germany).[19] Local toponyms were Polonised by the Polish Commission for the Determination of Place Names.[20]

Northern East Prussia to the Soviet Union

In April 1946, northern East Prussia became an official province of the Russian SFSR as "*Kenigsbergskaya Oblast*", with the Memel Territory becoming part of the Lithuanian SSR. In June 1946 114,070 German and 41,029 Soviet citizens were registered in the Oblast, with an unknown number of disregarded unregistered persons. In July of that year, the historic city of Königsberg was renamed Kaliningrad to honour Mikhail Kalinin and the area named the Kaliningrad Oblast. Between 24 August and 26 October 1948 21 transports with in total 42,094 Germans left the Oblast to the Soviet Occupation Zone. The last remaining Germans left in November 1949 (1,401 persons) and January 1950 (7 persons).[21] After the expulsion of the German population ethnic Russians, Belarusians, and Ukrainians were settled in the northern part.

In the Soviet part of the region, a policy of eliminating all remnants of German history was pursued. All German place names were replaced by new Russian names. The exclave was a military zone, which was closed to foreigners; Soviet citizens could only enter with special permission. In 1967 the remnants of Königsberg Castle were demolished on the orders of Leonid Brezhnev to make way for a new "House of the Soviets".

Königsberg Castle, 1895

Modern status

Since the fall of Communism in 1991, some German groups have tried to help settle the Volga Germans from eastern parts of Russia in the Kaliningrad Oblast. This effort was only a small success, however, as most impoverished Volga Germans preferred to emigrate to the richer Federal Republic of Germany, where they could become German citizens through the right of return.

Although the 1945–1949 expulsion of Germans from the northern part of former East Prussia was often conducted in a violent and aggressive way by Soviet officials seeking revenge for Nazi crimes committed in the Soviet Union, the present Russian inhabitants of the Kaliningrad Oblast have much less animosity towards Germans. German names have been revived in commercial Russian trade and there is sometimes

"House of the Soviets", built on the site of the former Königsberg Castle

talk of reverting Kaliningrad's name to its historic name of Königsberg. The city centre of Kaliningrad was completely rebuilt, as British bombs in 1944 and the Soviet siege in 1945 had left it in nothing but ruins.

The borders of the present-day Warmian-Masurian Voivodeship in Poland correspond closely to those of southern East Prussia.

See also

- Lithuania Minor
- List of cities and towns in East Prussia
- Drang nach Osten
- East Colonisation
- Kaliningrad Oblast
- Landsmannschaft Ostpreußen
- Masuria

Russian "Königsberg" licence plate, 2009

- Teutonic Knights
- Warmia

References

[1] The Columbia Encyclopedia, Sixth Edition (2008), East Prussia (http://www.encyclopedia.com/topic/East_Prussia.aspx)

[2] Ostpreußen: The Great Trek (http://www.exulanten.com/preussen.html)

[3] *Encyclopædia Britannica*: Old-Prussian-language (http://www.britannica.com/eb/article-9056977/Old-Prussian-language); Gordon, Raymond G., Jr. (ed.): Ethnologue: Languages of the World, 2005, Prussian (http://www.ethnologue.com/show_language.asp?code=prg)

[4] The Family Dönhoff, or the futility of revenge (http://www.ruf.rice.edu/~sarmatia/195/davies2.html)

[5] A Treatise on Political Economy (http://oll.libertyfund.org/index.php?option=com_staticxt&staticfile=show.php?title=274& chapter=38099&layout=html)

[6] The Prussians, "Ideal Prussians", Old Prussian and New Prussian (http://donelaitis.vdu.lt/prussian/princip.htm)

[7] Ethnic Groups and Population Changes in Twentieth-Century Central-Eastern Europe: History, Data, and Analysis. Piotr Eberhardt,page 166, 2003 M E Sharpe Inc

[8] Matull, Wilhelm (1973). "Ostdeutschlands Arbeiterbewegung: Abriß ihrer Geschichte, Leistung und Opfer" (http://library.fes.de/breslau/ pdf/a20715/a20715_06.pdf) (in German). Holzner Verlag. p. 350. .

[9] Die aufrechten Roten von Königsberg (http://einestages.spiegel.de/static/topicalbumbackground/4374/ die_aufrechten_roten_von_koenigsberg.html) Spiegel.de, 28 June 2009 (German)

[10] Matull, page 357

[11] Neumärker, Uwe et al (2007) (in German). *"Wolfsschanze": Hitlers Machtzentrale im Zweiten Weltkrieg* (3 ed.). Ch. Links Verlag. ISBN 3-86153-433-9.

[12] Szkolnictwo polskie w Niemczech 1919-1939, Henryk Chałupczak Wydawnictwo Uniwersytetu Marii Curie-Skłodowskiej,page9 1996

[13] Verwaltungsgeschichte.de (http://www.verwaltungsgeschichte.de/p_ostpreussen.html#einwohnerzahl)

[14] Beevor, Antony, *Berlin: The Downfall 1945*, chapters 1-8, Penguin Books (2002). ISBN 0-670-88695-5

[15] Beevor, Antony, *Berlin: The Downfall 1945*, Penguin Books (2002). ISBN 0-670-88695-5

[16] Andreas Kossert, Damals in Ostpreussen, p. 168, München 2008 ISBN 978-3-421-04366-5

[17] Ther, Philip; Siljak, Anna (2001). *Redrawing nations: ethnic cleansing in East-Central Europe, 1944-1948* (http://books.google.de/ books?id=oGmTs2SceAgC&dq=polonization+masuria&printsec=frontcover&source=bl&ots=H4rQU4mfaV& sig=lKf7dT3ypISTwxbcg1IOYQ_iekA&hl=de&ei=YOnVStz3KYeqsAbNob3bCw&sa=X&oi=book_result&ct=result&resnum=7& ved=0CBsQ6AEwBjha#v=onepage&q=masuria&f=false). Rowman&Littlefield Publishers. p. 109. ISBN 0-7425-1094-8. .

[18] Ethnic Germans in Poland and the Czech Republic:A Comparative Evaluation (http://www.stefanwolff.com/working-papers/ EthnicGermansPolandandCzechRepublic.pdf) by Karl Cordell and Stefan Wolff

[19] Andreas Kossert, Ostpreussen - Geschichte und Mythos, p.352, ISBN 3-88680-808-4

[20] The Polish toponymic guidelines (http://www.gugik.gov.pl/komisja/pliki/the_polish_toponymic_guidelines.pdf) (p.9)

[21] Andreas Kossert, Damals in Ostpreussen, p. 179-183, München 2008 ISBN 978-3-421-04366-5

Bibliography

Publications in English

- Baedeker, Karl, *Northern Germany*, 14th revised edition, London, 1904.
- Beevor, Antony (2002). "chapters 1-8" (http://www.antonybeevor.com/Berlin/berlinmenu.htm). *Berlin: The Downfall 1945*. Penguin Books. ISBN 0-670-88695-5. (on the years 1944/45)
- Alfred-Maurice de Zayas, " Nemesis at Potsdam". London, 1977. ISBN 0-8032-4910-1.
- Alfred-Maurice de Zayas, *A Terrible Revenge: The Ethnic Cleansing of the East European Germans, 1944-1950*, 1994, ISBN 0-312-12159-8
- Dickie, Reverend J.F., with E.Compton, *Germany*, A & C Black, London, 1912.
- von Treitschke, Heinrich, *History of Germany* - vol.1: *The Wars of Emancipation*, (translated by E & C Paul), Allen & Unwin, London, 1915.
- Powell, E. Alexander, *Embattled Borders*, London, 1928.
- Prausser, Steffen and Rees, Arfon: The Expulsion of the "German" Communities from Eastern Europe at the End of the Second World War. Florence, Italy, European University Institute, 2004.
- Naimark, Norman: Fires of Hatred. Ethnic Cleansing in Twentieth-Century Europe. Cambridge, Harvard University Press, 2001.
- Steed, Henry Wickham, *Vital Peace - A Study of Risks*, Constable & Co., London, 1936.
- Newman, Bernard, *Danger Spots of Europe*, London, 1938.
- Wieck, Michael: *A Childhood Under Hitler and Stalin: Memoirs of a "Certified Jew,"* University of Wisconsin Press, 2003, ISBN 0-299-18544-3.
- Woodward, E.L., Butler, Rohan; Medlicott, W.N., Dakin, Douglas, & Lambert, M.E., et al. (editors), *Documents on British Foreign Policy 1919-1939*, Three Series, Her Majesty's Stationery Office (HMSO), London, numerous volumes published over 25 years. Cover the Versailles Treaty including all secret meetings; plebiscites and all other problems in Europe; includes all diplomatic correspondence from all states.
- Previté-Orton, C.W., Professor, *The Shorter Cambridge Medieval History*, Cambridge University Press, 1952 (2 volumes).
- Balfour, Michael, and John Mair, *Four-Power Control in Germany and Austria 1945-1946*, Oxford University Press, 1956.
- Kopelev, Lev, *To Be Preserved Forever*, ("Хранить вечно"), 1976.
- Koch, H.W., Professor, *A History of Prussia*, Longman, London, 1978/1984, (P/B), ISBN 0-582-48190-2
- Koch, H.W., Professor, *A Constitutional History of Germany in the 19th and 20th Centuries*, Longman, London, 1984, (P/B), ISBN 0-582-49182-7
- MacDonogh, Giles, *Prussia*, Sinclair-Stevenson, London, 1994, ISBN 1-85619-267-9
- Nitsch, Gunter, *Weeds Like Us*, AuthorHouse, 2006, ISBN 978-1-4259-6755-0

Publications in German

- B. Schumacher: *Geschichte Ost- und Westpreussens*, Würzburg 1959
- Boockmann, Hartmut: *Ostpreußen und Westpreußen* (= Deutsche Geschichte im Osten Europas). Siedler, Berlin 1992, ISBN 3-88680-212-4
- Buxa, Werner and Hans-Ulrich Stamm: *Bilder aus Ostpreußen*
- Dönhoff, Marion Gräfin v. :*Namen die keiner mehr nennt - Ostpreußen, Menschen und Geschichte*
- Dönhoff, Marion Gräfin v.: *Kindheit in Ostpreussen*
- Falk, Lucy: *Ich Blieb in Königsberg. Tagebuchblätter aus dunklen Nachkriegsjahren*
- Kibelka, Ruth: *Ostpreußens Schicksaljahre, 1945-1948*
- Bernd, Martin (1998). *Masuren, Mythos und Geschichte*. Karlsruhe: Evangelische Akademie Baden. ISBN 83-85135-93-6.
- Nitsch, Gunter: "Eine lange Flucht aus Ostpreußen", Ellert & Richter Verlag, 2011, ISBN 978-3-8319-0438-9

- Wieck, Michael: *Zeugnis vom Untergang Königsbergs: Ein "Geltungsjude" berichtet,* Heidelberger Verlaganstalt, 1990, 1993, ISBN 3-89426-059-9.

Publications in French

- Pierre Benoît, *Axelle*
- Georges Blond, *L'Agonie de l'Allemagne*
- Michel Tournier, *Le Roi des aulnes*

Publications in Polish

- K. Piwarski (1946). *Dzieje Prus Wschodnich w czasach nowożytnych.* Gdańsk.
- Gerard Labuda, ed. (1969–2003). *"Historia Pomorza", vol. I–IV.* Poznań.
- collective work (1958–61). *"Szkice z dziejów Pomorza", vol. 1–3.* Warszawa.
- Andreas Kossert (2009). *PRUSY WSCHODNIE, Historia i mit.* Warszawa. ISBN 978-83-7383-354-8.

External links

- Brandenburg Prince-Electors co-inheritors 1568, co-regent 1577 (http://books.google.com/books?id=nduuOHX8Nl8C&pg=PA230&dq=mitbelehnung+preussen#PPA230,M1)
- East Prussia FAQ (http://users.foxvalley.net/~goertz/faqopr.html)
- East and West Prussia Gazetteer (http://www.progenealogists.com/germany/prussia/index.html)
- Provinz Ostpreußen (http://www.provinz-ostpreussen.de/home/index.html) (German)
- Ostpreußen.net (http://www.ostpreussen.net/) (German)
- Ostpreußen Info - East Prussia Information (http://www.ostpreussen-info.de/) (German)
- East- and West Prussia in Photos (http://www.ordensland.de)
- *Spuren der Vergangenheit / Следы Прошлого* (Traces of the past) (http://www.milovsky-gallery.albertina.ru/) This site by W.A. Milowskij, a Kaliningrad resident, contains hundreds of interesting photos, often with text explanations, of architectural and infrastructural artifacts of the territory's long German past. (German) (Russian)
- German Empire: Province of East Prussia (http://www.deutsche-schutzgebiete.de/provinz_ostpreussen.htm) (German)
- Britannica 2007 article (http://www.britannica.com/eb/article-9031793/East-Prussia)
- Growing up in East Prussia (http://www.jugendzeit-ostpreussen.de/konzept_en.html) An oral history project, documenting the German history of East Prussia with memories and reports by contemporary witnesses (German) (Polish)

Crown of the Kingdom of Poland

This article is about the unit of administrative division; for "public properties, state properties" of Polish-Lithuanian Kings see: Crown lands in Poland and Lithuania (Polish: królewszczyzny, dobra królewskie); for insignia of Polish and Polish-Lithuanian Kings see: Polish Crown Jewels

The **Crown of the Kingdom of Poland** (Polish: *Korona Królestwa Polskiego*, Latin: *Corona Regni Poloniae*), or simply **the Crown** (   ), is the name for the semi-legal concept related to the Polish monarchy, or even Poland itself.[1] In its political aspect, the concept of the Crown referred the Polish kingdom (nation) as distinctly separate from the person of the monarch.[1] It also had a geographical aspect, as it refferred to all lands that the Polish state (not the monarch) could claim to have the right to rule over.[1]

The Crown marked in   on high-level administration map of Polish-Lithuanian Commonwealth and its fiefdoms in 1619. Superimposed on present day political map of Central and Eastern Europe

History

The concept was introduced to Poland around 14th century, and was most likely transmitted from Hungary, and can be traced to England in the 12th century.[1] It was further strenghtened by the growing traditions of the royal elections in Poland, which helped to justify the idea that the nation is separate from the person of the king, and similar customs of parliamentary participation of nobility in the government (sejmik, great sejm).[1]

Political aspect

It marked a milestone in the evolution of Polish statehood, and represented the concept of the Polish kingdom (nation) as distinctly separate from the person of the monarch.[1] The introduction of that concept marked the transformation of the Polish government from the patrimonial monarchy to the class monarchy (*monarchia stanowa*).[1]

A related concept that evolved soon afterward was that of Rzeczpospolita, both of which served as the alternate names for the Polish state.[1] The Crown of Poland was also related to other symbols of Poland, such as the capital (Kraków), Polish coat of arms and the flag of Poland.[1]

Geographical aspect

The concept of Crown also had a geographical aspect, in particularly related to the indivisibility of the Polish (Crown) territory.[1] It can be also seen as a unit of administrative division, the territories under direct administration of Polish state from middle-ages to late 18th century (currently lands of Ukraine, Poland, some border lands of inter alia: Russia, Belarus, Moldova, Slovakia, Romania). Some of them belonged to the early Kingdom of Poland, then to Polish-Lithuanian Commonwealth until its final collapse in 1795.

At the same time, the Crown also referred to all lands that the Polish state (not the monarch) could claim to have the right to rule over, including those that were not within Polish borders.[1]

The term distinguishes those territories from federated with the Crown Grand Duchy of Lithuania (   ) and from fiefdom territories (which enjoyed varying degrees of autonomy or semi-independence from the King) inter alia the Duchy of Prussia (   ), the Duchy of Courland (   ).

Prior to the 1569 Union of Lublin, Crown territories may be understood as those of Poland proper, inhabited by Poles or other areas under the sovereignty of Polish nobility. With the Union of Lublin, however, most of present-day Ukraine (which had a negligible Polish population and had until then been governed by Lithuania) passed under Polish administration, becoming likewise Crown territory.

In that period, a term for a Pole was *koroniarz* (plural: *koroniarze*), derived from *Korona*.

> Depending on context, "Crown" may also refer to "The Crown," a term used to distinguish the personal influence and private assets of the Commonwealth's current monarch from government authority and property. This often meant a distinction between persons loyal to the elected King (royalists) and persons loyal to the magnates.

Provinces

Crown was divided into two provinces: Lesser Poland (Polish: Małopolska) and Greater Poland (Polish: Wielkopolska) which were further divided into administrative units known as voivodeships (Polish names of voivodships and towns below in brackets).

Greater Poland Province

- Brześć Kujawski Voivodeship (województwo brzesko-kujawskie, Brześć Kujawski)
- Gniezno Voivodeship (województwo gnieźnieńskie, Gniezno) from 1768
- Inowrocław Voivodeship (województwo inowrocławskie, Inowrocław)
- Kalisz Voivodeship (województwo kaliskie, Kalisz)
- Łęczyca Voivodeship (województwo łęczyckie, Łęczyca)
- Mazovian Voivodeship (województwo mazowieckie, of Mazowsze, Warsaw)
- Poznań Voivodeship (województwo poznańskie, Poznań)
- Płock Voivodeship (województwo płockie, Płock)
- Podlaskie Voivodeship (województwo podlaskie, Drohiczyn)

(Polish) Voivodeships of the Commonwealth of the Two Nations

- Rawa Voivodeship (województwo rawskie, Rawa)
- Sieradz Voivodeship (województwo sieradzkie, Sieradz)
- Prince-Bishopric of Warmia

Lesser Poland Province

- Bełz Voivodeship (województwo bełzkie, Bełz)
- Bracław Voivodeship (województwo bracławskie, Bracław)
- Czernichów Voivodeship (województwo czernichowskie, Czernichów)
- Kijów Voivodeship (województwo kijowskie, Kijów)
- Kraków Voivodeship (województwo krakowskie, Kraków)
- Lublin Voivodeship (województwo lubelskie, Lublin)
- Podole Voivodeship (województwo podolskie, Kamieniec Podolski)
- Ruś Voivodeship (województwo ruskie, Lwów), divided into
- Sandomierz Voivodeship (województwo sandomierskie, Sandomierz)
- Wołyń Voivodeship (województwo wołyńskie, Łuck)
- Duchy of Siewierz (Siewierz)

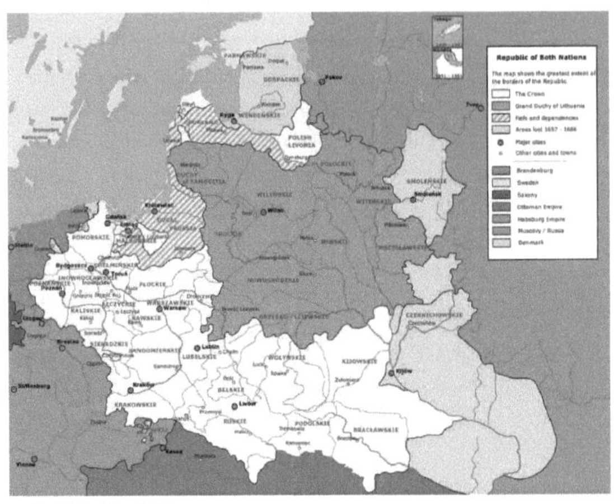

(Polish) (English) Map showing voivodeships of the Commonwealth of the Two Nations

Royal Prussia Province (1569 - 1772)

Royal Prussia Polish: *Prusy Królewskie*) was a province of the Kingdom of Poland from 1466 and then the Polish-Lithuanian Commonwealth from 1569 to 1772. Royal Prussia included Pomerelia, Chełmno Land (Kulmerland), Malbork Voivodeship (Marienburg), Gdańsk (Danzig), Toruń (Thorn), and Elbląg (Elbing).

Towns in Spisz County (1412 - 1795)

As one of the terms of the Treaty of Lubowla, the Hungarian crown exchanged, for a loan of *sixty times the amount of 37,000 Prague groschen* - approximately seven tonnes of pure silver, 16 rich salt-producing towns in the area of Spisz (Zips), as well as a right to incorporate them into Poland until the debt is repaid. The towns affected were: Biała, Lubica, Wierzbów, Spiska Sobota, Poprad, Straże, Spiskie Włochy, Nowa Wieś, Spiska Nowa Wieś, Ruszkinowce, Wielka, Spiskie Podgrodzie, Maciejowce, Twarożne.

References

[1] Juliusz Bardach, Boguslaw Lesnodorski, and Michal Pietrzak, *Historia panstwa i prawa polskiego* (Warsaw: Paristwowe Wydawnictwo Naukowe, 1987, p.85-86

See also

- Administrative division of the Polish-Lithuanian Commonwealth
- Lands of the Crown of Saint Stephen
- Lands of the Crown of Saint Wenceslaus

Kiwity, Warmian-Masurian Voivodeship

<table>
<tr><td colspan="2" align="center">Kiwity</td></tr>
<tr><td colspan="2" align="center">— Village —</td></tr>
<tr><td colspan="2" align="center">
Kiwity</td></tr>
<tr><td colspan="2" align="center">Coordinates: 54°6′0″N 20°46′14″E</td></tr>
<tr><td>Country</td><td>Poland</td></tr>
<tr><td>Voivodeship</td><td>Warmian-Masurian</td></tr>
<tr><td>County</td><td>Lidzbark</td></tr>
<tr><td>Gmina</td><td>Kiwity</td></tr>
<tr><td>Population</td><td>500</td></tr>
</table>

Kiwity [ki'vitɨ] (German *Kiewitten*) is a village in Lidzbark County, Warmian-Masurian Voivodeship, in northern Poland. It is the seat of the gmina (administrative district) called Gmina Kiwity.[1] It lies approximately 13 kilometres (**unknown operator: u'strong'** mi) east of Lidzbark Warmiński and 40 km (**unknown operator: u'strong'** mi) north-east of the regional capital Olsztyn.

Before 1772 the area was part of Kingdom of Poland, 1772-1945 Prussia and Germany (East Prussia).

The village has a population of 500.

References

[1] "Central Statistical Office (GUS) - TERYT (National Register of Territorial Land Apportionment Journal)" (http://www.stat.gov.pl/broker/access/prefile/listPreFiles.jspa) (in Polish). 2008-06-01. .

Olsztyn

<table>
<tr><td colspan="2" align="center">Olsztyn</td></tr>
<tr><td colspan="2" align="center">
Old Town</td></tr>
<tr><td colspan="2" align="center">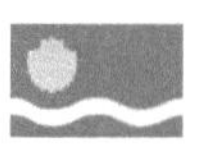 
Flag   Coat of arms</td></tr>
<tr><td colspan="2" align="center">Motto: Olsztyn – Miasto Młode Duchem…
(Olsztyn– city of a young spirit…)</td></tr>
<tr><td colspan="2" align="center">
Olsztyn
Coordinates: 53°47′N 20°30′E</td></tr>
<tr><td>Country</td><td>Poland</td></tr>
<tr><td>Voivodeship</td><td>Warmian-Masurian</td></tr>
<tr><td>County</td><td>city county</td></tr>
<tr><td>Established</td><td>14th century</td></tr>
<tr><td>Town rights</td><td>1353</td></tr>
<tr><td>Government</td><td></td></tr>
<tr><td>• Mayor</td><td>Piotr Grzymowicz</td></tr>
<tr><td>Area</td><td></td></tr>
</table>

• City	88.328 km^2 (**unknown operator: u'strong'** sq mi)
Highest elevation	154 m (**unknown operator: u'strong'** ft)
Lowest elevation	88 m (**unknown operator: u'strong'** ft)
Population (2009)	
• City	176387
• Density	**unknown operator: u'strong'**/km^2 (**unknown operator: u'strong'**/sq mi)
• Metro	270000
Time zone	CET (UTC+1)
• Summer (DST)	CEST (UTC+2)
Postal code	10-001 to 11–041
Area code(s)	+48 89
Car plates	NO
Website	http://www.olsztyn.eu

Olsztyn [ˈɔlʂtɨn] (listen) (German: *Allenstein* (listen); Old Polish: *Holstin*; Old Prussian: *Alnāsteini*; Lithuanian: *Olštynas*) is a city on the Łyna River in northeastern Poland. Olsztyn has been the capital of the Warmian-Masurian Voivodeship since 1999. It was previously in the Olsztyn Voivodeship (and in other units in 1945–75 and 1975–98). The city has county status.

Ordensburg castle built by the Teutonic Knights

History

In 1346 the old Prussian Warmian forest in the vicinity was cleared and a place was selected on the *Alle*, now Łyna River, for a new settlement. The Teutonic Knights began the construction of Ordensburg castle in 1347 as a stronghold against the Old Prussians, and the settlement of Allenstein was first mentioned the following year. The German name *Allenstein* meant a castle on the Alle River. It became known in Polish transliteration as *Olsztyn*. The settlement received municipal rights from Johannes von Leysen on 31 October 1353, and the castle was completed in 1397. Allenstein was incorporated into the Kingdom of Poland during the Polish-Lithuanian-Teutonic War in 1410 and in 1414 during the Hunger War, but was returned to the monastic state of the Teutonic Knights after hostilities ended.

Allenstein joined the Prussian Confederation in 1440. It rebelled against the Teutonic Knights in 1454 upon the outbreak of the Thirteen Years' War and requested protection from the Polish Crown. Although the Teutonic Knights captured the town in the following year, it was retaken by Polish troops in 1463. The Second Peace of Thorn (1466) allocated Allenstein and the Bishopric of Warmia as part of Royal Prussia under the sovereignty of the Crown of Poland. From 1516–21, Nicolaus Copernicus lived at the castle as administrator of Allenstein and Mehlsack (Pieniężno); he was in charge of the defenses of Allenstein and Warmia during the Polish-Teutonic War of 1519–21.

Allenstein was sacked by Swedish troops in 1655 and 1708 during the Polish-Swedish wars, and the town's population was nearly wiped out in 1710 by epidemics of bubonic plague and cholera.

Allenstein was annexed by the Kingdom of Prussia in 1772 during the First Partition of Poland. A Prussian census recorded a population of 1,770 people, predominantly farmers, in Allenstein, which was administered within the Province of East Prussia. It was visited by Napoleon Bonaparte in 1807 after his victories over the Prussian Army at

Jena and Auerstedt. In 1825 the city was inhabited by 1266 Poles and 1341 Germans[1] The German language newspaper, *Allensteiner Zeitung*, was first published in 1841. The town hospital was founded in 1867.

Allenstein's Kopernikusplatz (now Plac Bema) in 1917

Allenstein became part of the German Empire in 1871 during the Prussian-led unification of Germany. Two years later the city was connected by railway to Thorn (Toruń). Its first Polish language newspaper, the *Gazeta Olsztyńska*, was founded in 1886. Allenstein's infrastructure developed rapidly: gas was installed in 1890, telephones in 1892, public water supply in 1898, and electricity in 1907. The city became the capital of Regierungsbezirk Allenstein, a government administrative region in East Prussia, in 1905. From 1818–1910 the city was administered within the East Prussia Allenstein District, after which it became an independent city.

Shortly after the outbreak of World War I, troops of the Russian Empire captured Allenstein in 1914, but it was recovered by the Imperial German Army in the Battle of Tannenberg. The battle actually took place much closer to Allenstein than Tannenberg (now Stębark), but the victorious Germans, having been defeated in the medieval battle of Tannenberg, named it as such for propaganda purposes). In 1920 during the East Prussian plebiscite, Allenstein voted to remain in German East Prussia instead of becoming part of the Second Polish Republic. The football club SV Hindenburg Allenstein played in Allenstein from 1921–45. After the Nazi seizure of power in 1933, Poles and Jews in Allenstein were increasingly persecuted. In 1935 the Wehrmacht made the city the seat of the *Allenstein Militärische Bereich*. It was the home of the 11th Infanterie Division, the 11th Artillery Regiment, and the 217th Infanterie Division.

In 1920 a plebiscite was held to determine whether the city's populace wished to remain in East Prussia or become part of Poland. In order to advertise the plebiscite, special postage stamps were produced by overprinting German stamps and sold from 3 April. One kind of overprint read **PLÉBISCITE / OLSZTYN / ALLENSTEIN**, while the other read **TRAITÉ / DE / VERSAILLES / ART. 94 et 95** inside an oval whose border gave the full name of the plebiscite commission. Each overprint was applied to 14 denominations ranging from 5 Pf to 3 M.

5-Pfennig stamp

The plebiscite was held on 11 July, and produced 362,209 votes (97.8 %) for East Prussia and 7,980 votes (2.2 %) for Poland. The stamps became invalid on 20 August. Despite their short period of use, almost all of the stamps are cheaply available both used and unused.

On 12 October 1939, after the invasion of Poland that began World War II, the Wehrmacht established an Area Headquarters for Wehrkreis I that controlled the sub-areas of Allenstein, Lötzen (Giżycko) and Zichenau (Ciechanów). Beginning in 1939, members of the Polish-speaking minority, especially members of the Union of Poles in Germany, were deported to Nazi concentration camps.

Allenstein was plundered and burnt by the invading Soviet Red Army on 22 January 1945, as the Eastern Front reached the city. Allenstein's German population evacuated the region or were subsequently expelled. On 2 August 1945, the city was placed under Polish administration by the Soviets (according to the Potsdam Agreement) and officially renamed to Polish *Olsztyn*. In October 1945, the German population of Olsztyn was expelled by Order of the City Commanders of Olsztyn and ordered to assemble on 18 October at Karl Roensch Street barracks camp for transfer to Germany and in the case of non-compliance were to be put in a "punishment camp".

A tyre factory was founded in Olsztyn in 1967.

Geography

Located in the north-east part of Poland in the "Thousands Lakes Area"

Green belt

More than half of the forests occupying 21.2 % of the city area form a single complex of the Municipal Forest (1050 ha) used mainly for recreation and tourism purposes. Within the Municipal Forest area are situated two peat-land flora sanctuaries, Mszar and Redykajny. Municipal greenery (560 ha, 6.5 % of the town area) developed in the form of numerous parks, green spots and three cemeteries over a century-old. The greenery includes 910 monuments of nature and groups of protected trees in the form of beech, oak, maple and lime-lined avenues.

Lakes

The city is situated in a lake region of forests and plains. There are 15 lakes inside the administrative bounds of the city (13 with areas greater than 1 ha). The overall area of lakes in Olsztyn is about 725 ha, which constitutes 8.25 % of the total city area.

Lake	Area (ha)	Maximum depth (m)
Lake Ukiel (a.k.a. Jezioro Krzywe)	412	43
Lake Kortowskie	89.7	17.2
Lake Track (a.k.a. Jezioro Trackie)	52.8	4.6
Lake Skanda	51.5	12
Lake Redykajny	29.9	20.6
Lake Długie	26.8	17.2
Lake Sukiel	20.8	25
Lake Tyrsko (a.k.a. Jezioro Gutkowskie)	18.6	30.6
Lake Stary Dwór (a.k.a. Jezioro Starodworskie)	6.0	23.3
Lake Siginek (a.k.a. Jezioro Kopytko, Jezioro Podkówka, Jezioro Styginek)	6.0	insufficient data
Lake Czarne	approximately 1.3	insufficient data
Lake Żbik	approximately 1.2	insufficient data
Lake Pereszkowo (a.k.a. Jezioro Pyszkowo)	approximately 1.2	insufficient data
Lake Mummel (a.k.a. Jezioro Mumel)	approximately 0.3	insufficient data
Lake Modrzewiowe	0.25	insufficient data

Demographics

The Upper Gate (High Gate) in the Old Town

Year	Population
1772	1,770
1846	4,000
1875	6,000
1885	11,555
1890	19,373
1895	25,000
1939	50,000
1941	54,300
1946	23,000
1950	45,000
1972	over 100,000
1994	165,000
2000	170,000
2009 (30 June)	176,387

Administrative division

Olsztyn is divided into 23 districts:

Fish Market

City hall

District	Population	Area	Density
Brzeziny	1,456	2.25 km²	647.1/km²
Dajtki	5,863	7.5 km²	781.7/km²
Generałów	6,500	no data	no data
Grunwaldzkie	6,027	1.46 km²	4,128.1/km²
Gutkowo	2,256	7.2 km²	313.3/km²
Jaroty	29,046	4.82 km²	6,026.1/km²
Kętrzyńskiego	7,621	4.83 km²	1,577.8/km²
Kormoran	16,166	1.1 km²	14,696.4/km²
Kortowo	1,131	4.22 km²	268/km²
Kościuszki	6,704	1.18 km²	5,681.4/km²
Likusy	2,286	2.1 km²	1,088.6/km²
Mazurskie	4,615	5.98 km²	771.7/km²
Nad Jeziorem Długim	2,408	4.23 km²	569.3/km²
Nagórki	12,538	1.69 km²	7,418.9/km²
Pieczewo	10,918	2.24 km²	4,874.1/km²
Podgrodzie	11,080	1.35 km²	8,207.4/km²
Podleśna	10,414	9.93 km²	1,048.7/km²

Pojezierze	13,001	2.39 km²	5,439.7/km²
Redykajny	1,555	6.1 km²	254.9/km²
Śródmieście	3,448	0.58 km²	5,944.8/km²
Wojska Polskiego	6,759	5.03 km²	1,343.7/km²
Zatorze	6,988	0.45 km²	15,528.9/km²
Zielona Górka	1,015	6.44 km²	157.6/km²

There are many smaller districts: Jakubowo, Karolin, Kolonia Jaroty, Kortowo II, Łupstych, Niedźwiedź, Piękna Góra, Podlesie, Pozorty, Skarbówka Poszmanówka, Słoneczny Stok, Stare Kieźliny, Stare Miasto, Stare Zalbki, Stary Dwór, Track. These do not have council representative assemblies.

Culture

Theatres

- Stefan Jaracz Theatre (est. 1925)
- Puppet Theatre

Cinemas

- Helios

Museums

Olsztyn's largest museum is the Museum of Warmia and Mazury. The city also has the Gazeta Olsztyńska House, Museum of Nature, and Museum of Sports.

Architecture

- The Old Town
- The Gothic castle of the Bishopric of Warmia built during the 14th century.
- St. James's Cathedral (Polish: *św. Jakuba*, German: *St. Jacob* or *St. Jakob*).
- Old Town Hall on the Market Square – built in mid-14th century.
- Gazeta Olsztyńska House at Fish Market.
- The town walls and the Upper Gate (since the mid-19th century known as the High Gate).
- Neogothic church of the Holy Heart of Jesus, built during the years 1901–1902
- The New City Hall

St. James's Cathedral

- The Railway Bridge over the River Łyna gorge near Artyleryjska and Wyzwolenia streets, built during the years 1872–1873
- The Jerusalem Chapel, built in 1565
- Church of St. Lawrence, built during the late 14th century
- FM- and TV-mast Olsztyn-Pieczewo – 360 metres high, since the collapse of the Warsaw radio mast the tallest structure in Poland

The history of the Castle in Olsztyn:

The castle was built between 1346–1353 and by then it had one wing on the north-east side of the rectangular courtyard. Access to the castle, lead from the drawbridge of the river Łyna (Alle), surrounded by a belt of defensive walls and a moat. The south-west wing of the castle was built in the 15th century, tower situated in the west corner of the courtyard, from the middle of 14th century, was rebuilt in the early 16th century and had a round shape on a square base and was 40 meters high. At the same time the castle walls were raised to a height of 12 meters and a second belt of the lower walls was built. The castle walls were partly combined with city walls, which made a castle looks like it had been a powerful bastion defending the access to the city. The castle was owned by Warmia Chapter, which until 1454 together with the Prince-Bishop of Warmia, was under military protection of the Teutonic Knights and their Monastic State of Prussia. The castle played a huge role in the Polish-Teutonic wars by then. After the Battle of Grunwald in 1410, the Poles took it after a few days siege. In the Thirteen Years' War (1454–66) it was jumping from hands to hands. The Knights threatened the castle and the town even in 1521, but the defense was very effective. They confined of the one, failed assault. Not many people know that with the history of the castle and the city of Olsztyn is connected with Nicolaus Copernicus. He prepared the defense of Olsztyn against the invasion of the Teutonic Knights. In the sixteenth century, there were two prince-bishops of Warmia that has stayed there: Johannes Dantiscus – "the first sarmatian poet, endowed with imperial laurel wreath for" Latin Songs "(1538, 1541) and Marcin Kromer, who formed with equal ease in Latin and Polish scientific and literary works (1580). Kromer consecrated the chapel of St. Anna, which was built in the south-west wing of the castle. In the course of time both wings of the castle lost military importance, which for residential purposes has become very convenient. In 1779 Prince-Bishop Ignacy Krasicki stopped here as well. After the Royal Prussian annexation of Warmia in 1772, the castle became the property of the state board of estates (War and Domain Chamber, Kriegs- und Domänenkammer). In 1845 the bridge over the moat was replaced by a dam connecting the castle with the city even better than before, therefore was dried. In 1901–1911 the general renovation of the castle was performed, however, several sections of the building were violated at the same time which they changed the original look of the castle e.g. putting on window frames in a cloister. The tower that was crowned in 1921 and again in 1926 in the halls of the castle, became a museum. The whole castle is a museum until today. In 1945 it became a residence of the Masurian Museum, which today is called the Museum of Warmia and Masuria. In addition to all that and the exhibition activities in Olsztyn, there are also popular events held within the frameworks of the Olsztyn Artistic Summer and so called "evenings of the castle" and "Sundays in the Museum".

Music

Death metal act Vader, regarded as one of the first Death metal bands from Poland.

Economy

The Michelin tyre company (former Stomil Olsztyn) is the largest employer in the region of Warmia and Masuria.[2] Other important industries are food processing and furniture manufacturing.

Unemployment

Olsztyn- 4,000 (people) – 4.8% rate

Education

* University of Warmia and Mazury [3]
* University of Computer Science and Economics [4]
* Olsztyńska Szkoła Wyższa im. Józefa Rusieckiego [5]
* Olsztyńska Wyższa Szkoła Informatyki i Zarządzania im. Tadeusza Kotarbińskiego [6]
* Wyższe Seminarium Duchowne HOSIANUM [7]
* Masurian Institute (est. 1943)

University of Warmia and Mazury

Sport

* Indykpol AZS Olsztyn – men's volleyball team playing in Polish Volleyball League (PLS, Polska Liga Siatkówki)
* OKS 1945 Olsztyn – men's football team, (8 seasons in Polish Ekstraklasa as Stomil Olsztyn)
* Warmia Traveland Olsztyn – men's handball team playing in Seria A (Polish First League)
* AZS UWM Trójeczka Olsztyn – men's basketball team playing in Polish Second League
* WMPD Olsztyn – men's rugby team, playing in First Polish League
* Budowlani Olsztyn Wrestling team

Politics

Members of the Sejm elected from Olsztyn constituency in 2005

* Mieczysław Aszkiełowicz, Self-Defense of the Republic of Poland (Samoobrona Rzeczypospolitej Polskiej)
* Beata Bublewicz, Civic Platform (PO, Platforma Obywatelska)
* Jerzy Gosiewski, Law and Justice (PiS, Prawo i Sprawiedliwość)
* Tadeusz Iwiński, Democratic Left Alliance (SLD, Sojusz Lewicy Demokratycznej)
* Edward Ośko, League of Polish Families (LPR, Liga Polskich Rodzin)
* Adam Puza, Law and Justice (PiS, Prawo i Sprawiedliwość)
* Sławomir Rybicki, Civic Platform (PO, Platforma Obywatelska)
* Lidia Staroń, Civic Platform (PO, Platforma Obywatelska)
* Aleksander Marek Szczygło, Law and Justice (PiS, Prawo i Sprawiedliwość)
* Zbigniew Włodkowski, Polish Peasant Party (PSL, Polskie Stronnictwo Ludowe)

Members of Senate elected from Olsztyn constituency in 2005

* Ryszard Józef Górecki, Civic Platform (PO, Platforma Obywatelska)
* Jerzy Szmit, Law and Justice (PiS, Prawo i Sprawiedliwość)

Notable residents

- Johannes von Leysen (1310–1388), town founder
- Nicolaus Copernicus (1473–1543), astronomer, administrator, and town commander
- Johannes Knolleisen (+1511), German academic and provider of academic stipends
- Lucas David (1503–1583), German historian of Prussia
- Karl Roensch (1858–1921), industrialist, city governmental official
- Hugo Haase (1863–1919), German politician, jurist and pacifist
- Franz Justus Rarkowski (1873–1950), military bishop (1938–45)
- August Trunz (1875–1963), founder of the Prussica-Sammlung Trunz

Statue of Nicolaus Copernicus in front of the castle

- Hubert Hönnekes (1880–1947), teacher at Kopernikus-Schule, member of East Prussian Landtag (provincial assembly), member of the Gernman Reichstag (national parliament) 1930–1933
- Erich Mendelsohn (1887–1953), German-Jewish architect who fled the Nazis
- Olga Desmond (1891–1964), dancer and actress
- Kurd von Bülow (1899–1971), geologist
- Klaus-Joachim Zülch (1910–1988), neuro scientist
- Günter Wand (1912–2002), conductor
- Georg Hermanowski (1918–1993), author, translator
- Hans-Jürgen Wischnewski (1922–2005), politician
- Curt Lowens (*1925), German actor
- Leonhard Pohl (*1929), gymnast
- Maximilian Kaller from 1930–1947 Bishop of Ermland
- Wolfgang Milde (*1934), library director, handwriting specialist
- Wolf Lepenies (*1941), sociologist, political scientist and author
- Georg Schimanski, camera man
- Kurt Baluses soccer player and trainer of VFB Stuttgart
- Ulrich Schrade (1943–2009), German-Polish philosopher and pedagogue
- Marian Bublewicz (1950–1993), Polish rally and racing driver of the 80s and 90s
- Juliusz Machulski (*1955), director
- Izabela Trojanowska (*1955), actress and singer
- Krzysztof Hołowczyc (*1962), rally driver
- Artur Wojdat (*1968), swimmer
- Józef Glemp (*1929), from 1979–1981 Bishop of Warmia, seat at Olsztyn
- Wojciech Grzyb (*1981), volleyball player
- Tomasz Zahorski soccer player
- Piotr "Peter" Wiwczarek (*1965), guitarist and vocalist, frontman for the death metal band Vader.

International relations

Twin towns – Sister cities

Olsztyn is twinned with:

Planetarium of Olsztyn

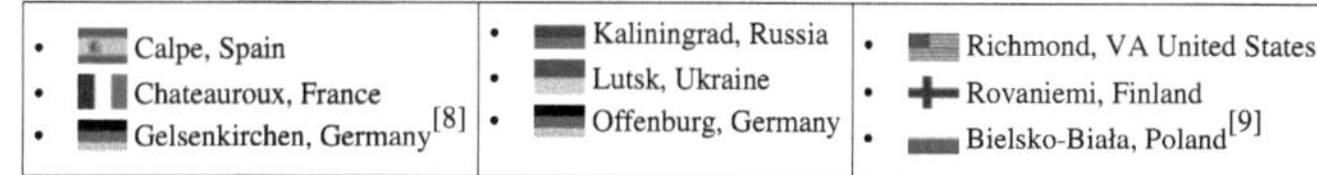

• Calpe, Spain	• Kaliningrad, Russia	• Richmond, VA United States
• Chateauroux, France	• Lutsk, Ukraine	• Rovaniemi, Finland
• Gelsenkirchen, Germany[8]	• Offenburg, Germany	• Bielsko-Biała, Poland[9]

Olsztyn belongs to the Federation of Copernicus Cities, an association of cities where Copernicus lived and worked, such as Bologna, Frombork, Kraków, and Toruń. The main office of the federation is situated at Olsztyn Planetarium and Astronomical Observatory, located on St. Andrew's Hill (143 m) in a former water tower erected in 1897.

References

[1] Historia Pomorza:(1815-1850),Gerard Labuda, Poznańskie Towarzystwo Przyjaciół Nauk,page 157, 1993

[2] "100-milionowa opona w olsztyńskiej fabryce Michelin – Autoflesz.pl – Niezależny Portal Motoryzacyjny" (http://www.autoflesz.pl/artykuly/4718,100milionowa_opona_w_olsztynskiej_fabryce_Michelin.html). Autoflesz.pl. . Retrieved 16 September 2011.

[3] http://www.uwm.edu.pl/pl/strona.php?b=2&ids=1

[4] http://wsiie.olsztyn.pl/english/

[5] http://www.osw.olsztyn.pl/

[6] http://www.owsiiz.edu.pl/

[7] http://www.hosianum.edu.pl/

[8] "List of Twin Towns in the Ruhr District" (http://www.twins2010.com/fileadmin/user_upload/pic/Dokumente/List_of_Twin_Towns_01.pdf?PHPSESSID=2edd34819db21e450d3bb625549ce4fd). 2009 Twins2010.com (http://www.twins2010.com/index.php?id=home&L=1). . Retrieved 28 October 2009.

[9] "Bielsko-Biała – Partner Cities" (http://www.um.bielsko.pl/). 2008 Urzędu Miejskiego w Bielsku-Białej.. . Retrieved 10 December 2008.

• (Polish) http://www.olsztyn.eu/

• (Polish) http://www.bezrobocie.net/stat_powiaty.php/

• (http://www.zamkigotyckie.org.pl/olsztyn.htm) (polish)

• (http://www.pascal.pl/atrakcja.php?id=25797) (polish)

External links

- Official website (http://www.olsztyn.eu/en.html) (English)
- Olsztyn City Guide (http://www.olsztyn.com/) (English)

Lidzbark Warmiński

<table>
<tr><td colspan="2" align="center">Lidzbark Warmiński</td></tr>
<tr><td colspan="2" align="center">
Warmian Bishop's castle</td></tr>
<tr><td colspan="2">
Coat of arms</td></tr>
<tr><td colspan="2" align="center">
Lidzbark Warmiński</td></tr>
<tr><td colspan="2" align="center">Coordinates: 54°7′N 20°35′E</td></tr>
<tr><td>Country</td><td>Poland</td></tr>
<tr><td>Voivodeship</td><td>Warmian-Masurian</td></tr>
<tr><td>County</td><td>Lidzbark</td></tr>
<tr><td>Gmina</td><td>Lidzbark Warmiński (urban gmina)</td></tr>
</table>

Established	before 1240
Town rights	1308
Government	
• **Mayor**	Artur Wajs
Area	
• **Total**	14.34 km^2 (**unknown operator: u'strong'** sq mi)
Population (2006)	
• **Total**	16390
• **Density**	**unknown operator: u'strong'**/km^2 (**unknown operator: u'strong'**/sq mi)
Time zone	CET (UTC+1)
• **Summer (DST)**	CEST (UTC+2)
Postal code	11-100 to 11-102
Area code(s)	+48 89
Car plates	NLI
Website	[1]

Lidzbark Warmiński ['lʲid ˈzbarɡ var'mʲiɲskʲi] (◀ listen) (German: *Heilsberg* (◀ listen)) is a town in the Warmian-Masurian Voivodeship in Poland. It is the capital of Lidzbark County.

History

The town was originally an Old Prussian settlement known as *Lecbarg* until being conquered in 1240 by the Teutonic Knights, who called it *Heilsberg*. In 1306 it became the seat for the Bishopric of Warmia and remained the Prince-Bishop's seat for 500 years. In 1309 the settlement received town privileges. After the Second Peace of Thorn (1466), the town was integrated into the Polish province of Royal Prussia.

Nicolaus Copernicus lived at the castle for several years, and it is believed he wrote part of his *De revolutionibus orbium coelestium* there.

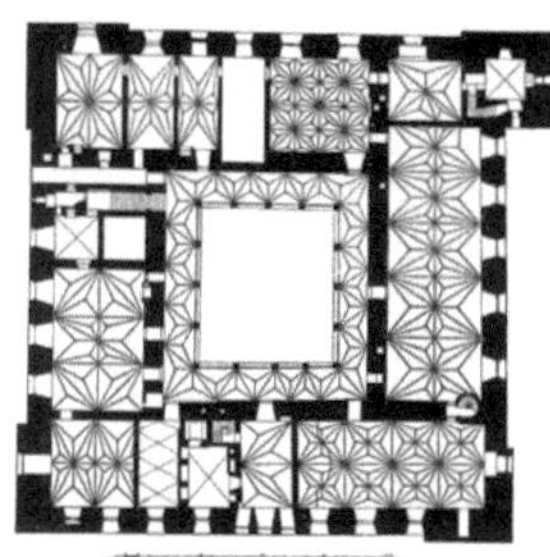

Castle floor plan

In the winter of 1703-04 the town was the residence of King Charles XII of Sweden during the Great Northern War.

Lidzbark was annexed with the rest of Warmia by the Kingdom of Prussia in the First Partition of Poland in 1772. In 1807 a battle took place near the town between the French under Murat and Soult and the Russians and Prussians under Bennigsen.

From 1933-45 Heilsberg was the site of the large radio station Transmitter Heilsberg. The town was heavily damaged during its conquest by the Soviet Red Army during World War II in 1945. Afterwards, the town was ceded to Poland and its ethnic German population was expelled to the west.

Education

- Wszechnica Warmińska, founded on 21 May 2004
- Comprehensive Schools im. Kazimierza Jagiellończyka
- Trade Schools im. Stanisława Staszica
- Farner School
- Primary School No. 1 im. Mikołaja Kopernika
- Primary School No. 3 im. Ignacego Krasickiego
- Primary School No. 4 im. Jana Pawła II
- High/grammar School No. 1
- High/grammar School No. 2
- National Music School I Level
- Not Public Kindergarten No. "Kubuś"
- Not Public Kindergarten No. "Miś"
- Not Public Kindergarten No. "Puchatek"
- Public Kindergarten No. 5
- Public Kindergarten No. 6

Famous Persons

- Nicolaus Copernicus (Mikołaj Kopernik) (1473–1547), famous astronomer, mathematician, physician, and canon
- Eustathius von Knobelsdorf (1519–71), Warmia Domherr, poet, lyricist of noble family von Knobelsdorf
- Matthias Johann Meyer or Matthias Meyer, (died 1737) baroque painter
- Ernst Burchard (1876–1920), doctor and scientist
- Bruno Hippler (1894–1942), officer
- Ferdinand Schulz (1892–1929), German sail gliding pioneer, named Ikarus von Ostpreussen
- Zbigniew Mikołejko (born 1951), philosopher and historian of religion

External links

- (Polish) http://www.lidzbark.com/
- (Polish) http://lidzbarkwarminski.pl/
- (Polish) http://www.warmia-mazury.pl/powiaty/lidzbark.html
- Map of Warmia Catholic Diocese in 1755 [2]
- (German) http://www.heilsberg.org/

References

[1] http://www.lidzbarkwarminski.pl
[2] http://www.domwarminski.pl/images/stories/warmia_regionem/mapy_historyczne_tabula_geografica_w.jpg

Lidzbark Warmiński

<table>
<tr><td colspan="2" align="center">Lidzbark Warmiński</td></tr>
<tr><td colspan="2" align="center">

Warmian Bishop's castle</td></tr>
<tr><td colspan="2">

Coat of arms</td></tr>
<tr><td colspan="2" align="center">

Lidzbark Warmiński</td></tr>
<tr><td colspan="2" align="center">Coordinates: 54°7′N 20°35′E</td></tr>
<tr><td>Country</td><td>Poland</td></tr>
<tr><td>Voivodeship</td><td>Warmian-Masurian</td></tr>
<tr><td>County</td><td>Lidzbark</td></tr>
<tr><td>Gmina</td><td>Lidzbark Warmiński (urban gmina)</td></tr>
<tr><td>Established</td><td>before 1240</td></tr>
<tr><td>Town rights</td><td>1308</td></tr>
<tr><td>Government</td><td></td></tr>
<tr><td>• Mayor</td><td>Artur Wajs</td></tr>
</table>

Area	
• **Total**	14.34 km^2 (**unknown operator: u'strong'** sq mi)
Population (2006)	
• **Total**	16390
• **Density**	**unknown operator: u'strong'/km^2** (**unknown operator: u'strong'/sq mi**)
Time zone	CET (UTC+1)
• **Summer (DST)**	CEST (UTC+2)
Postal code	11-100 to 11-102
Area code(s)	+48 89
Car plates	NLI
Website	[1]

Lidzbark Warmiński [ˈlʲid ˆzbarg varˈmʲiɲskʲi] (◄ listen) (German: *Heilsberg* (◄ listen)) is a town in the Warmian-Masurian Voivodeship in Poland. It is the capital of Lidzbark County.

History

The town was originally an Old Prussian settlement known as *Lecbarg* until being conquered in 1240 by the Teutonic Knights, who called it *Heilsberg*. In 1306 it became the seat for the Bishopric of Warmia and remained the Prince-Bishop's seat for 500 years. In 1309 the settlement received town privileges. After the Second Peace of Thorn (1466), the town was integrated into the Polish province of Royal Prussia.

Nicolaus Copernicus lived at the castle for several years, and it is believed he wrote part of his *De revolutionibus orbium coelestium* there.

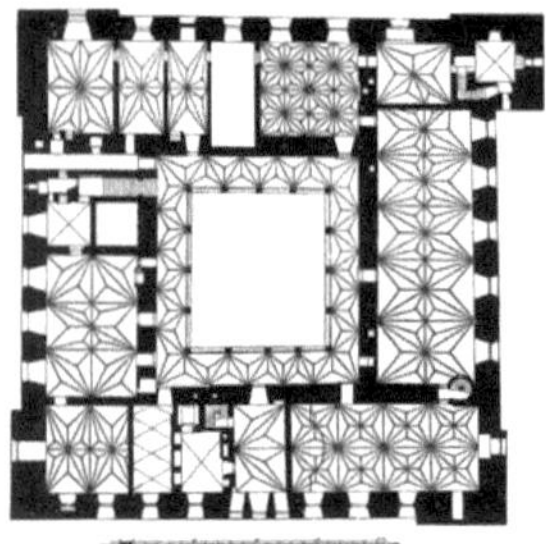
Castle floor plan

In the winter of 1703-04 the town was the residence of King Charles XII of Sweden during the Great Northern War.

Lidzbark was annexed with the rest of Warmia by the Kingdom of Prussia in the First Partition of Poland in 1772. In 1807 a battle took place near the town between the French under Murat and Soult and the Russians and Prussians under Bennigsen.

From 1933-45 Heilsberg was the site of the large radio station Transmitter Heilsberg. The town was heavily damaged during its conquest by the Soviet Red Army during World War II in 1945. Afterwards, the town was ceded to Poland and its ethnic German population was expelled to the west.

Education

- Wszechnica Warmińska, founded on 21 May 2004
- Comprehensive Schools im. Kazimierza Jagiellończyka
- Trade Schools im. Stanisława Staszica
- Farner School
- Primary School No. 1 im. Mikołaja Kopernika
- Primary School No. 3 im. Ignacego Krasickiego
- Primary School No. 4 im. Jana Pawła II
- High/grammar School No. 1
- High/grammar School No. 2
- National Music School I Level
- Not Public Kindergarten No. "Kubuś"
- Not Public Kindergarten No. "Miś"
- Not Public Kindergarten No. "Puchatek"
- Public Kindergarten No. 5
- Public Kindergarten No. 6

Famous Persons

- Nicolaus Copernicus (Mikołaj Kopernik) (1473–1547), famous astronomer, mathematician, physician, and canon
- Eustathius von Knobelsdorf (1519–71), Warmia Domherr, poet, lyricist of noble family von Knobelsdorf
- Matthias Johann Meyer or Matthias Meyer, (died 1737) baroque painter
- Ernst Burchard (1876–1920), doctor and scientist
- Bruno Hippler (1894–1942), officer
- Ferdinand Schulz (1892–1929), German sail gliding pioneer, named Ikarus von Ostpreussen
- Zbigniew Mikołejko (born 1951), philosopher and historian of religion

External links

- (Polish) http://www.lidzbark.com/
- (Polish) http://lidzbarkwarminski.pl/
- (Polish) http://www.warmia-mazury.pl/powiaty/lidzbark.html
- Map of Warmia Catholic Diocese in 1755 [2]
- (German) http://www.heilsberg.org/

Lidzbark County

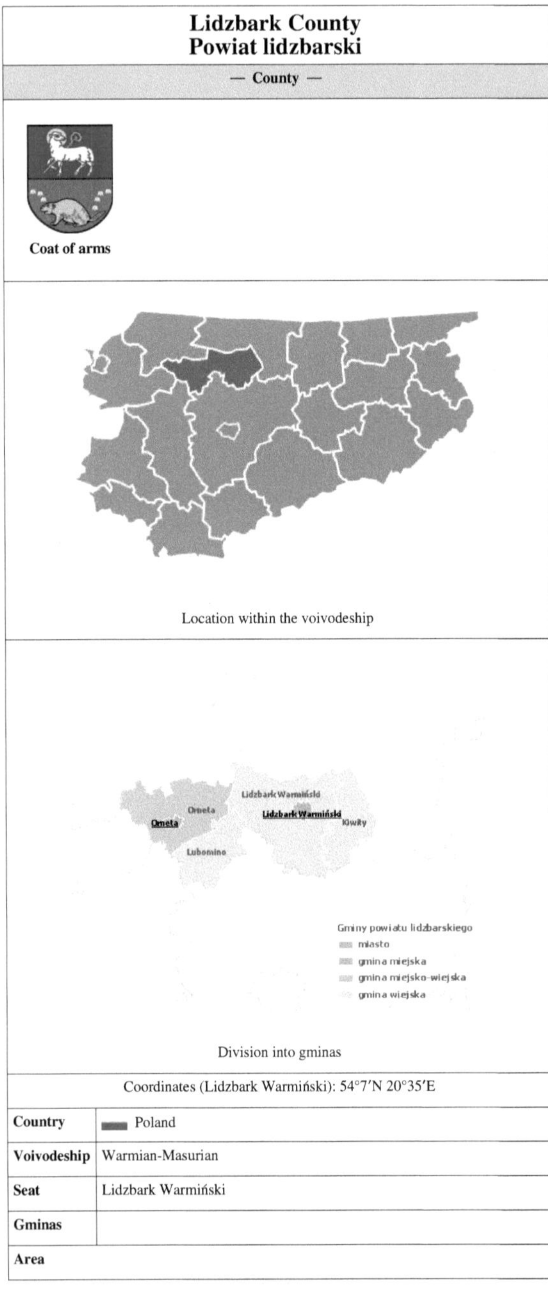

<table>
<tr><td colspan="2" align="center">Lidzbark County
Powiat lidzbarski</td></tr>
<tr><td colspan="2" align="center">— County —</td></tr>
<tr><td colspan="2" align="center">Coat of arms</td></tr>
<tr><td colspan="2" align="center">Location within the voivodeship</td></tr>
<tr><td colspan="2" align="center">Division into gminas</td></tr>
<tr><td colspan="2" align="center">Coordinates (Lidzbark Warmiński): 54°7′N 20°35′E</td></tr>
<tr><td>Country</td><td>Poland</td></tr>
<tr><td>Voivodeship</td><td>Warmian-Masurian</td></tr>
<tr><td>Seat</td><td>Lidzbark Warmiński</td></tr>
<tr><td>Gminas</td><td></td></tr>
<tr><td>Area</td><td></td></tr>
</table>

• **Total**	924.42 km^2 (**unknown operator: u'strong'** sq mi)
Population (2006)	
• **Total**	43006
• **Density**	**unknown operator: u'strong'**/km^2 (**unknown operator: u'strong'**/sq mi)
• **Urban**	25770
• **Rural**	17236
Car plates	NLI
Website	[1]

Lidzbark County (Polish: *powiat lidzbarski*) is a unit of territorial administration and local government (powiat) in Warmian-Masurian Voivodeship, northern Poland. It came into being on January 1, 1999, as a result of the Polish local government reforms passed in 1998. Its administrative seat and largest town is Lidzbark Warmiński, which lies 38 kilometres (**unknown operator: u'strong'** mi) north of the regional capital Olsztyn. The only other town in the county is Orneta, lying 30 km (**unknown operator: u'strong'** mi) west of Lidzbark Warmiński.

The county covers an area of 924.42 square kilometres (**unknown operator: u'strong'** sq mi). As of 2006 its total population is 43,006, out of which the population of Lidzbark Warmiński is 16,390, that of Orneta is 9,380, and the rural population is 17,236.

Neighbouring counties

Lidzbark County is bordered by Bartoszyce County to the north-east, Olsztyn County to the south, Ostróda County to the south-west, and Elbląg County and Braniewo County to the west.

Administrative division

The county is subdivided into five gminas (one urban, one urban-rural and three rural). These are listed in the following table, in descending order of population.

Gmina	Type	Area (km²)	Population (2006)	Seat
Lidzbark Warmiński	urban	14.3	16,390	
Gmina Orneta	urban-rural	244.1	12,701	Orneta
Gmina Lidzbark Warmiński	rural	371.0	6,733	Lidzbark Warmiński *
Gmina Lubomino	rural	149.6	3,717	Lubomino
Gmina Kiwity	rural	145.4	3,465	Kiwity
				* seat not part of the gmina

References

- Polish official population figures 2006 [2]

References

[1] http://www.powiatlidzbarski.pl/
[2] http://www.stat.gov.pl/gus/45_655_PLK_HTML.htm

Gmina Kiwity

<table>
<tr><td colspan="2" align="center">Gmina Kiwity
Kiwity Commune</td></tr>
<tr><td colspan="2" align="center">— Gmina —</td></tr>
<tr><td colspan="2" align="center">Coordinates (Kiwity): 54°6′0″N 20°46′14″E</td></tr>
<tr><td>Country</td><td>Poland</td></tr>
<tr><td>Voivodeship</td><td>Warmian-Masurian</td></tr>
<tr><td>County</td><td>Lidzbark</td></tr>
<tr><td>Seat</td><td>Kiwity</td></tr>
<tr><td>Area</td><td></td></tr>
<tr><td>• Total</td><td>145.38 km^2 (unknown operator: u'strong' sq mi)</td></tr>
<tr><td>Population (2006)</td><td></td></tr>
<tr><td>• Total</td><td>3465</td></tr>
<tr><td>• Density</td><td>unknown operator: u'strong'/km^2 (unknown operator: u'strong'/sq mi)</td></tr>
</table>

Gmina Kiwity is a rural gmina (administrative district) in Lidzbark County, Warmian-Masurian Voivodeship, in northern Poland. Its seat is the village of Kiwity, which lies approximately 13 kilometres (**unknown operator: u'strong'** mi) east of Lidzbark Warmiński and 40 km (**unknown operator: u'strong'** mi) north-east of the regional capital Olsztyn.

The gmina covers an area of 145.38 square kilometres (**unknown operator: u'strong'** sq mi), and as of 2006 its total population is 3,465.

Villages

Gmina Kiwity contains the villages and settlements of Bartniki, Czarny Kierz, Kiersnowo, Kierwiny, Kiwity, Klejdyty, Klutajny, Kobiela, Konity, Krekole, Maków, Napraty, Połapin, Rokitnik, Samolubie, Tolniki Wielkie and Żegoty.

Neighbouring gminas

Gmina Kiwity is bordered by the gminas of Bartoszyce, Bisztynek, Jeziorany and Lidzbark Warmiński.

References

- Polish official population figures 2006 [2]

Village

A **village** is a clustered human settlement or community, larger than a hamlet with the population ranging from a few hundred to a few thousand (sometimes tens of thousands), Though often located in rural areas, the term urban village is also applied to certain urban neighbourhoods, such as the East Village in Manhattan, New York City and the Saifi Village in Beirut, Lebanon, as well as Hampstead Village in the London conurbation. Villages are normally permanent, with fixed dwellings; however, transient villages[1] can occur. Further, the dwellings of a village are fairly close to one another, not scattered broadly over the landscape, as a dispersed settlement.

Berber village in Ourika valley, High Atlas, Morocco.

Historically, villages were a usual form of community for societies that practise subsistence agriculture, and also for some non-agricultural societies. In Great Britain, a hamlet earned the right to be called a village when it built a church.[2] In many cultures, towns and cities were few, with only a small proportion of the population living in them. The Industrial Revolution attracted people in larger numbers to work in mills and factories; the concentration of people caused many villages to grow into towns and cities. This also enabled specialization of labor and crafts, and development of many trades. The trend of urbanization continues, though not always in connection with industrialisation. Villages have been eclipsed in importance as units of human society and settlement.

Bangladeshi Village

Traditional villages

Although many patterns of village life have existed, the typical village was small, consisting of perhaps 5 to 30 families. Homes were situated together for sociability and defence, and land surrounding the living quarters was farmed. Traditional fishing villages were based on artisan fishing and located adjacent to fishing grounds.

Masouleh village, Gilan Province, Iran

An alpine village in the Lötschental Valley, Switzerland

South Asia

India "The soul of India lives in its villages", declared M. K. Gandhi[3] at the beginning of 20th century. According to the 2011 census of India, 68.84% of Indians (around 833.1 million people) live in 640,867 different villages.[4] The size of these villages varies considerably. 236,004 Indian villages have a population less than 500, while 3,976 villages have a population of 10,000+. Most of the villages have their own temple, mosque or church depending on the local religious following.

A village in central India.

East Asia

People's Republic of China In mainland China, villages are divisions under township Zh:or town Zh:.

Republic of China (Taiwan) In the Republic of China (Taiwan), villages are divisions under townships or county-controlled cities. The village is called a *tsuen* or *cūn* () under a rural township () and a *li* () under an urban township () or a county-controlled city. See also Li (unit).

Japan

South Korea

Shirakawa-gō, Nagano Japan.

Southeast Asia

Thailand

Brunei, Indonesia, Malaysia and Singapore

In Indonesia, depending on the principles they are administered, villages are called *desa* or *kelurahan*. A *desa* (a term that derives from a Sanskrit word meaning "country" that is found in a name such as "Bangladesh") is administered according to traditions and customary law (*adat*), while a *kelurahan* is administered along more "modern" principles. *Desa* are generally located in rural areas while *kelurahan* are generally urban subdivisions. A village head is respectively called *kepala desa* or *lurah*. Both are elected by the local community. A *desa* or *kelurahan* is itself the subdivision of a *kecamatan* (district), in turn the subdivision of a *kabupaten* (regency).

The *nagari* of Pariangan, West Sumatra.

The same general concept applies all over Indonesia. How ever, there is some variation among the vast numbers of Austronesian ethnic groups. For instance, in Bali villages have been created by grouping traditional hamlets or *banjar*, which constitute the basis of Balinese social life. In the Minangkabau country in West Sumatra province traditional villages are called *nagari* (a term deriving from another Sanskrit word meaning "city", which can be found in a name like "Srinagar"). In some areas such as Tanah Toraja, elders take turns watching over the village at a command post. As a general rule, *desa* and *kelurahan* are groupings of hamlets (*kampung* in Indonesian, *dusun* in the

Javanese language, *banjar* in Bali).

In Malaysia, the term *kampung* (sometimes spelling *kampong*) in the English language has been defined specifically as "a Malay hamlet or village in a Malay-speaking country".[5] In other words, a **kampung** is defined today as a village in Brunei, Indonesia, Singapore, and Malaysia. In Malaysia, a *kampung* is determined as a locality with 10,000 or fewer people. Since historical times, every Malay village came under the leadership of a *penghulu* (village chief), who has the power to hear civil matters in his village (see Courts of Malaysia for more details). A Malay village typically contains a *"masjid"* (mosque) or *"surau"* (Muslim chapel), paddy fields and Malay houses on stilts. Malay and Indonesian villagers practice the culture of helping one another as a community, which is better known as "joint bearing of burdens" (*gotong royong*),[6] as well as being family-oriented (especially the concept of respecting one's family [particularly the parents and elders]), courtesy and believing in God (*"Tuhan"*) as paramount to everything else. It is common to see a cemetery near the mosque, as all Muslims in the Malay or Indonesian village want to be prayed for, and to receive Allah's blessings in the afterlife. In Sarawak, some villages are called 'long' pronounced as 'long' in Chinese. These villages are mostly found in western Sarawak.

Singapore also follows the Malaysian *kampung*. However, there are only a few *kampung* villages remaining, mostly on islands surrounding Singapore such as Pulau Ubin. In the past, there was many *kampung* villages in Singapore but now there aren't many on the mainland.

Philippines

In urban areas of the Philippines, the term "village" most commonly refers to private subdivisions, especially gated communities. These villages emerged in the mid-20th century and were initially the domain of elite urban dwellers. Those are common in major cities in the country and their residents have a wide range of income levels. Such villages may or may not correspond to administrative units (usually barangays) and/or be privately administered. Barangays more correspond to the villages of old times, and the chairman (formerly a village datu) now settles intrapersonal matters or polices the village, though with much less authority and respect than in Indonesia or Malaysia.

Vietnam

Village, or "làng", is a basis of Vietnam society. Vietnam's village is the typical symbol of Asian agricultural production. Vietnam's village typically contains: a village gate, "lũy tre" (bamboo hedges), "đình làng" (communal house) where "thành hoàng" (tutelary god) is worshiped, a common well, "đồng lúa" (rice field), "chùa" (temple) and houses of all families in the village. All the people in Vietnam's villages usually have a blood relationship. They are farmers who grow rice and have the same traditional handicraft. Vietnam's villages have an important role in society (Vietnamese saying: "Custom rules the law" -"Phép vua thua lệ làng" [literally: the king's law yields to village customs]). Everyone in Vietnam wants to be buried in their village when they die.

Central and Eastern Europe

Slavic countries

Selo (Cyrillic: село; Polish: *wieś, sioło*) is a Slavic word meaning "village" in Bosnia and Herzegovina, Bulgaria, Croatia, Macedonia, Russia, Serbia, and Ukraine. For example there are numerous *sela* (plural of *selo*) called Novo Selo in Bulgaria, Croatia, Montenegro and others in Serbia, and Macedonia. In Slovenia, the word *selo* is used for very small villages (less than a thousand people) and in dialects; the Slovene word *vas* is used all over Slovenia.

Bulgaria

In Bulgaria, the different types of *Sela* vary from a small selo of 5 to 30 families to one of several thousand people. According to a 2002 census, in that year there were 2,385,000 Bulgarian citizens living in settlements classified as *villages*.[7] A 2004 Human Settlement Profile on Bulgaria[8] conducted by the United Nations Department of Economic and Social Affairs stated that:

Kovachevitsa, a village in southern Bulgaria

> The most intensive is the migration "city – city". Approximately 46% of all migrated people have changed their residence from one city to another. The share of the migration processes "village – city" is significantly less – 23% and "city – village" – 20%. The migration "village – village" in 2002 is 11%.[7]

It also stated that

> the state of the environment in the small towns and villages is good apart from the low level of infrastructure.[7]

In Bulgaria, it is becoming popular to visit villages for the atmosphere, culture, crafts, hospitality of the people and the surrounding nature. This is called *selski tourism* (Bulgarian: селски туризъм), meaning "village tourism".

Russia

In Russia, as of the 2010 Census, 26.3% of the country's population lives in rural localities;[9] down from 26.7% recorded in the 2002 Census.[9] Multiple types of rural localities exist, but the two most common are *derevnya* (деревня) and *selo* (село). Historically, the formal indication of status was religious: a city (*gorod*) had a cathedral, a *selo* had a church, while a *derevnya* had neither.

The lowest administrative unit of the Russian Empire, a *volost*, or its Soviet or modern Russian successor, a *selsoviet*, was typically headquartered in a *selo* and embraced a few neighboring villages.

The village of Zaponorye in Moscow Oblast

Between 1926 and 1989, Russia's rural population shrank from 76 million people to 39 million, due to urbanization, collectivization, dekulakization, and the World War II losses, but has nearly stabilized since. During 1930–1937, mass starvation in Russia and other parts of the Soviet Union lead to the death of at least 14.5 million peasants (including 5-7 million in the Holodomor).[10]

Most Russian rural localities have populations of less than 200 people, and the smaller places take the brunt of depopulation: e.g., in 1959, about one half of Russia's rural population lived in villages of fewer than 500 people, while now less than one third does. In the 1960s–1970s, the depopulation of the smaller villages was driven by the central planners' drive to get the farm workers out of smaller, "prospect-less" hamlets and into the collective or state farms' main villages, with more amenities.[11]

Most Russian rural residents are involved in agricultural work, and it is very common for villagers to produce their own food. As prosperous urbanites purchase village houses for their second homes, Russian villages sometimes are transformed into dacha settlements, used mostly for seasonal residence.

The historically Cossack regions of Southern Russia and parts of Ukraine, with their fertile soil and absence of serfdom, had a rather different pattern of settlement from central and northern Russia. While peasants of central Russia lived in a village around the lord's manor, a Cossack family often lived on its own farm, called *khutor*. A number of such *khutors* plus a central village made up the administrative unit with a center in a *stanitsa* (Russian: станица; Ukrainian: станиця, *stanytsia*). Such *stanitsas*often with a few thousand residents, were usually larger than a typical *selo* in central Russia.

The term *aul*/*aal* is used to refer mostly Muslim-populated villages in Caucasus and Idel-Ural, without regard to the number of residents.

Ukraine

In Ukraine, a village, known locally as a *"selo"* (село), is considered the lowest administrative unit. Villages may have an individual administration (*silrada*) or a joint administration, combining two or more villages. Villages may also be under the jurisdiction of a city council (*miskrada*) or town council (*selyshchna rada*) administration.

There is, however, another smaller type of settlement which is designated in Ukrainian as a *selysche* (селище). This type of community is generally referred to in English as a "settlement". In comparison with an urban-type settlement, Ukrainian legislation does not have a concrete definition or a criterion to differentiate such settlements from villages. They represent a type of a small rural locality that might have once been a *khutir*, a fisherman's settlement, or a dacha. They are administered by a *silrada* (council) located in a nearby adjacent village. Sometimes the term *"selysche"* is also used in a more general way to refer to adjacent settlements near a bigger city, including urban-type settlements (*selysche miskoho typu*) and/or villages; however, ambiguity is often avoided in connection with urbanized settlements by referring to them using the three-letter abbreviation *smt* instead.

The *khutir* (хутір) and *stanytsia* (станиця) are not part of the administrative division any longer, primarily due to collectivization. *Khutirs* were very small rural localities consisting of just few housing units and were sort of individual farms. They became really popular during the Stolypin reform in the early 20th century. During the collectivization, however, residents of such settlements were usually declared to be kulaks and had all their property confiscated and distributed to others (nationalized) without any compensation. The *stanitsa* likewise has not survived as an administrative term. The *stanitsa* was a type of a collective community that could include one or more settlements such as villages, *khutirs*, and others. Today, *stanitsa*-type formations have only survived in Kuban (Russian Federation) where Ukrainians were resettled during the time of the Russian Empire.

Western & Southern Europe

United Kingdom

A village in the UK is a compact settlement of houses, smaller in size than a town, and generally based on agriculture or, in some areas, mining (such as Ouston, County Durham), quarrying or sea fishing.

The major factors in the type of settlement are location of water sources, organisation of agriculture and landholding, and likelihood of flooding. For example, in areas such as the Lincolnshire Wolds, the villages are often found along the spring line halfway down the hillsides, and originate as spring line settlements, with the original open field systems around the village. In northern Scotland, most villages are planned to a grid pattern located on or close to major roads, whereas in areas such as the Forest of Arden, woodland clearances produced small hamlets around village greens.[12] [13]

The main street of the village of Castle Combe, Wiltshire, England.

Some villages have disappeared (for example, deserted medieval villages), sometimes leaving behind a church or manor house and sometimes nothing but bumps in the fields.Some show archaeological evidence of settlement at three or four different layers, each distinct from the previous one. Clearances may have been to accommodate sheep or game estates, or enclosure, or may have resulted from depopulation, such as after the Black Death or following a move of the inhabitants to more prosperous districts. Other villages have grown and merged and often form hubs within the general mass of suburbia — such as Hampstead, London and Didsbury in Manchester. Many villages are now predominantly dormitory locations and have suffered the loss of shops, churches and other facilities.

For many British people, the village represents an ideal of Great Britain. Seen as being far from the bustle of modern life, it is represented as quiet and harmonious, if a little inward-looking. This concept of an unspoilt Arcadia is present in many popular representations of the village such as the radio serial *The Archers* or the best kept village competitions. The reality is that many villages are plagued by lack of access to public transport and local services, specially affecting the poor and elderly who cannot afford their own means of transport.[14]

Many villages in South Yorkshire, North Nottinghamshire, North East Derbyshire, County Durham, South Wales and Northumberland are known as pit villages. These (such as Murton, County Durham) grew from hamlets when the sinking of a colliery in the early 20th century resulted in a rapid growth in their population and the colliery owners built new housing, shops, pubs and churches. Some pit villages outgrew nearby towns by area and population; for example, Rossington in South Yorkshire came to have over four times more people than the nearby town of Bawtry. Some pit villages grew to become towns; for example, Maltby in South Yorkshire grew from 600 people in the 19th century[15] to over 17,000 in 2007.[16] Maltby was constructed under the auspices of the Sheepbridge Coal and Iron Company and included ample open spaces and provision for gardens.[17]

Bisley, Gloucestershire, a village in the Cotswolds

In the UK, the main historical distinction between a hamlet and a village was that the latter had a church,[2] and so usually was the centre of worship for an ecclesiastical parish. However, some civil parishes may contain more than one village. The typical village had a pub or inn, shops, and a blacksmith. But many of these facilities are now gone, and many villages are dormitories for commuters. The population of such settlements ranges from a few hundred people to around five thousand. A village is distinguished from a town in that:

- A village should not have a regular agricultural market, although today such markets are uncommon even in settlements which clearly are towns.
- A village does not have a town hall nor a mayor.
- If a village is the principal settlement of a civil parish, then any administrative body that administers it at parish level should be called a parish council or parish meeting, and not a town council or city council. However, some civil parishes have no functioning parish, town, or city council nor a functioning parish meeting. In Wales, where the equivalent of an English civil parish is called a Community, the body that administers it is called a Community Council. However, larger councils may elect to call themselves town councils.[18] Unlike Wales, Scottish community councils have no statutory powers.[19]
- There should be a clear green belt or open fields, as, for example, seen on aerial maps for Ouston surrounding its parish[20] borders. However this may not be applicable to urbanised villages: although these may not considered to be villages, they are often widely referred to as being so; an example of this is Horsforth in Leeds.

France

Same general definition as in the UK.

An independent association named *Les Plus Beaux Villages de France*, was created in 1982 to promote assets of small and picturesque French villages of quality heritage. As of 2008, 152 villages in France have been listed in "The Most Beautiful Villages of France".

Spain

Spain has plenty of little villages around its territory. The concept of village and country life is really present and usual in the North of the country (Atlantic area), especially in Galicia where villages are similar to English ones.

South of Barcelona is Spain's most romantic Mediterranean beach town, with a 2.5 km-long (1 1/2-mile) sandy beach and a promenade studded with flowers and palm trees. Sitges is a town with a rich connection to art; Picasso and Dalí both spent time here.[21] Mérida is an important Roman town with great tapas. Barcena Mayor (Cantabria) has houses that date back to the sixth century with simple two floors

Saint-Cirq-Lapopie (Lot) is one of "The Most Beautiful Villages in France".

constructions and rectangular form. Salamanca, an ancient Celtic town, is also a Renaissance city with striking architecture. Its sandstones buildings have a beautiful lustre giving the city the nickname, La Ciudad Dorado.[22]

Morella, Castellón is a medieval village located in the region of "Comunitat Valenciana" with huge castles with a rich renaissance history.[23] Rogueira pasturelands is one of the great ecological jewels of Galicia. Rivers, pools and springs abound in this verdant forest, as do underground water caves and caverns with a prehistoric past.[24] San Marti Vell is a charming small village well known for its Gothic spire. La Bisbal should be next on the list. The town is worth visiting for its Main Square and the castle. This Romanesque castle is situated in the middle of the town, giving it a romantic look. There is also Palafrugell, Palau-sator, Sant Julia and Sant Feliu de Boada. They are all very important because of their medieval patrimony. Castelló d'Empúries has 13th century Gothic churches. Angles also possesses outstanding medieval constructions throughout its village.[25] All villages have a church or hermitage.

Portugal

Villages are more usual in the northern and central regions and in the Alentejo. Most of them have a church and a "Casa do Povo" (people's house), where the village's summer **romarias** or religious festivities are usually held. Summer is also when many villages are host to a range of folk festivals and fairs, taking advantage of the fact that many of the locals who reside abroad tend to come back to their native village for the holidays.

Netherlands

In the flood prone districts of the Netherlands, villages were traditionally built on low man-made hills called terps before the introduction of regional dyke-systems. In modern days, the term *dorp* (lit. "village") is usually applied to settlements no larger than 20,000, though there's no official law regarding status of settlements in the Netherlands.

Middle East

Lebanon

Like France, villages in Lebanon are usually located in remote mountainous areas. The majority of villages in Lebanon retain their Aramaic names or are derivative of the Aramaic names, and this is because Aramaic was still in use in Mount Lebanon up to the 18th century.[26]

The main square of Saifi Village in Centre Ville, Beirut, Lebanon

Many of the Lebanese villages are a part of districts, these districts are known as "kadaa" which includes the districts of Baabda (Baabda), Aley (Aley), Matn (Jdeideh), Keserwan (Jounieh), Chouf (Beiteddine), Jbeil (Byblos), Tripoli (Tripoli), Zgharta (Zgharta / Ehden), Bsharri (Bsharri), Batroun (Batroun), Koura (Amioun), Miniyeh-Danniyeh (Minyeh / Sir Ed-Danniyeh), Zahle (Zahle), Rashaya (Rashaya), Western Beqaa (Jebjennine / Saghbine), Sidon (Sidon), Jezzine (Jezzine), Tyre (Tyre), Nabatiyeh (Nabatiyeh), Marjeyoun (Marjeyoun), Hasbaya (Hasbaya), Bint Jbeil (Bint Jbeil), Baalbek (Baalbek), and Hermel (Hermel).

The district of Danniyeh consists of thirty six small villages, which includes Almrah, Kfirchlan, Kfirhbab, Hakel al Azimah, Siir, Bakhoun, Miryata, Assoun, Sfiiri, Kharnoub, Katteen, Kfirhabou, Zghartegrein, Ein Qibil.

Danniyeh (known also as Addinniyeh, Al Dinniyeh, Al Danniyeh, Arabic: الضنية سير) is a region located in Miniyeh-Danniyeh District in the North Governorate of Lebanon. The region lies east of Tripoli, extends north as far as Akkar District, south to Bsharri District and Zgharta District and as far east as Baalbek and Hermel. Dinniyeh has an excellent ecological environment filled with woodlands, orchards and groves. Several villages are located in this mountainous area, the largest town being Sir Al Dinniyeh.

An example of a typical mountainous Lebanese village in Dannieh would be Hakel al Azimah which is a small village that belongs to the district of Danniyeh, situated between Bakhoun and Assoun's boundaries. It is in the centre of the valleys that lie between the Arbeen Mountains and the Khanzouh.

Syria

Syria contains a large number of villages that vary in size and importance, including the ancient, historical and religious villages, such as Ma'loula, Sednaya, and Brad (Mar Maroun's time). The diversity of the Syrian environments creates significant differences between the Syrian villages in terms of the economic activity and the method of adoption. Villages in the south of Syria (Huran, Jabal Al-Arab), the north-east (the Syrian island) and the Orontes River

General view from Al-Annaze village, near Tartus, Syria

basin depend mostly on agriculture, mainly grain, vegetables and fruits. Villages in the region of Damascus and Aleppo depend on trading. Some other villages, such as Marmarita depend heavily on tourist activity.

Mediterranean cities in Syria, such as Tartus and Latakia have similar types of villages. Mainly, villages were built in very good sites which had the fundamentals of the rural life, like water. An example of a Mediterranean Syrian village in Tartus would be Al-Annaze, which is a small village that belongs to the area of Al Sauda. The area of Al Sauda is called a nahiya, which is a subdistrict.

Australasia & Oceania

Pacific Islands Communities on pacific islands were historically called villages by English speakers who traveled and settled in the area. Some communities such as several Villages of Guam continue to be called villages despite having large populations that can exceed 40,000 residents.

New Zealand The traditional Māori village was the pā, a fortified hill-top settlement. Tree-fern logs and flax were the main building materials. As in Australia (see below) the term is now used mainly in respect of shopping or other planned areas.

The village of Puamau on Hiva Oa, Marquesas Islands, French Polynesia

Australia The term village often is used in reference to small planned communities such as retirement communities or shopping districts, and tourist areas such as ski resorts. Small rural communities are usually known as townships. Larger settlements are known as towns.

South America

Argentina Usually set in remote mountainous areas, some also cater to winter sports and/or tourism, see: Uspallata, La Cumbrecita, Villa Traful and La Cumbre

North America

Canada

United States

Incorporated villages

In twenty[27] U.S. states, the term "village" refers to a specific form of incorporated municipal government, similar to a city but with less authority and geographic scope. However, this is a generality; in many states, there are villages that are an order of magnitude larger than the smallest cities in the state. The distinction is not necessarily based on population, but on the relative powers granted to the different types of municipalities and correspondingly, different obligations to provide specific services to residents.

A Newfoundland fishing village

In some states such as New York, Wisconsin, or Michigan, a village is an incorporated municipality, usually, but not always, within a single town or civil township. Residents pay taxes to the village and town or township and may vote in elections for both as well. In some cases, the village may be coterminous with the town or township. There are also many villages which span the boundaries of more than one town or township, and some villages may even straddle county borders.

There is no limit to the population of a village in New York; Hempstead, the largest village in the state, has 55,000 residents, making it more populous than some of the state's cities. However, villages in the state may not exceed five square miles (13 km²) in area.

In the state of Wisconsin, a village is always legally separate from the towns that it has been incorporated from. The largest village is Menomonee Falls, which has over 32,000 residents.

Michigan and Illinois also have no set population limit for villages and there are many villages that are larger than cities in those states. The village of Arlington Heights, IL had 75,101 residents as of the 2010 census.

Villages in Ohio are often legally part of the township from which they were incorporated, although exceptions such as Hiram exist, in which the village is separate from the township.[28] They have no area limitations, but become cities if they grow a population of more than 5,000.[29]

In Maryland, a locality designated "Village of ..." may be either an incorporated town or a special tax district.[30] An example of the latter is the Village of Friendship Heights.

In states that have New England towns, a "village" is a center of population or trade, including the town center, in an otherwise sparsely-developed town or city — for instance, the village of Hyannis in the city of the Barnstable, Massachusetts.

Unincorporated villages

In many states, the term "village" is used to refer to a relatively small unincorporated community, similar to a hamlet in New York state. This informal usage may be found even in states that have villages as an incorporated municipality, although such usage might be considered incorrect and confusing.

See also

* Global village
* Linear village
* Village green
* Village lock-up
* police village

Settlement types

* Dugout
* Fishing village
* Hamlet
* Microtown

Countries and localities

* Dhani and villages
* Dogon villages
* Hakka architecture
* Ksar
* List of villages in Europe by country
* Pueblo
* Sołectwo (rough equivalent in Poland)
* Ville

Developed environments

* Developed environments
* City
* Exurban
* Megalopolis
* Rural
* Suburban
* Urban area

Footnotes

[1] http://www.google.co.uk/search?hl=en&safe=off&q=%22transient+villages%22&btnG=Search&meta=

[2] Dr Greg Stevenson, "What is a Village?" (http://www.bbc.co.uk/history/programmes/restoration/2006/exploring_brit_villages_01.
     shtml), *Exploring British Villages*, BBC, 2006, accessed 20 October 2009

[3] http://www.pibbng.kar.nic.in/feature1.pdf

[4] "Indian Census" (http://www.censusindia.gov.in/). Censusindia.gov.in. . Retrieved 2012-04-09.

[5] "Meriam-Webster Online" (http://www.m-w.com/dictionary/kampung). M-w.com. 2007-04-25. . Retrieved 2010-03-28.

[6] Geertz, Clifford. "Local Knowledge: Fact and Law in Comparative Perspective", pp. 167-234 in Geertz *Local Knowledge: Further Essays in
     Interpretive Anthropology*, NY: Basic Books. 1983.

[7] "Human Settlement Country Profile, Bulgaria (*2004*)" (http://www.un.org/esa/agenda21/natlinfo/countr/bulgaria/
     Bulgariahumansettlement2003.PDF) (PDF). United Nations Department of Economic and Social Affairs. . Retrieved 2008-11-30.

[8] http://www.un.org/esa/agenda21/natlinfo/countr/bulgaria/Bulgariahumansettlement2003.PDF

[9] Федеральная служба государственной статистики (Federal State Statistics Service) (2011). "Предварительные итоги Всероссийской
     переписи населения 2010 года[[Category:Articles containing non-English language text (http://www.perepis-2010.ru/
     results_of_the_census/results-inform.php)] (*Preliminary results of the 2010 All-Russian Population Census*)"] (in Russian). *Всероссийская
     перепись населения 2010 года (2010 All-Russia Population Census)*. Federal State Statistics Service. . Retrieved February 9, 2012.

[10] Robert Conquest (1986) *The Harvest of Sorrow: Soviet Collectivization and the Terror-Famine*. Oxford University Press. ISBN
      0-19-505180-7.

[11] "Российское село в демографическом измерении" (*Rural Russia measured demographically*) (http://demoscope.ru/weekly/2006/0253/tema04.php) (Russian). This article reports the following census statistics:

Census year	1959	1970	1979	1989	2002
Total number of rural localities in Russia	294,059	216,845	177,047	152,922	155,289
Of them, with population 1 to 10 persons	41,493	25,895	23,855	30,170	47,089
Of them, with population 11 to 200 persons	186,437	132,515	105,112	80,663	68,807

[12] Wild, Martin Trevor (2004). *Village England: a social history of the countryside* (http://books.google.co.uk/books?id=M7AmyuGr5Y8C&printsec=frontcover). I.B.Tauris. p. 12. ISBN 978-1-86064-939-4. .

[13] Taylor, Christopher (1984). *Village and farmstead: A history of rural settlement in England* (http://books.google.co.uk/books?ei=NksHTpXqIo6t8QPE1ozADQ). G. Philip. p. 192. ISBN 978-0-540-01082-0. .

[14] OECD (2011). *OECD Rural Policy Reviews: England, United Kingdom 2011* (http://books.google.co.uk/books?id=bQCGWKfXJNMC&printsec=frontcover&source=gbs_ge_summary_r&cad=0). OECD Publishing. p. 237. .

[15] *The Parliamentary gazetteer of England and Wales* (http://books.google.co.uk/books?id=mxIQAAAAYAAJ&printsec=frontcover). **3**. A. Fullarton & Co.. 1851. p. 344. .

[16] "Maltby Ward" (http://www.rotherham.gov.uk/download/553/maltby_ward). Rotherham Metropolitan Borough Council. . Retrieved 2011-06-26.

[17] Baylies, Carolyn Louise (1993). *The history of the Yorkshire miners, 1881-1918* (http://books.google.co.uk/books?id=WEIOAAAAQAAJ&printsec=frontcover&source=gbs_ge_summary_r&cad=0). Routledge. .

[18] "National Statistics" (http://www.statistics.gov.uk/geography/parishes.asp). Statistics.gov.uk. . Retrieved 2010-03-28.

[19] "Portobello Community Council" (http://www.porty.org.uk/council/index.php). Porty.org.uk. . Retrieved 2010-03-28.

[20] "Ouston Parish Council" (http://parishes.durham.gov.uk/ouston/Pages/wherewelive.aspx). durham.gov.uk. .

[21] http://www.frommers.com/destinations/spain/0242020855.html

[22] http://www.tourclare.com/spanishtownsandvillages.php

[23] http://www.travelthruspain.com/what-to-do/cities

[24] http://www.spain.info/no/reportajes/sierra_de_o_courel_naturaleza_y_aldeas_medievales.html

[25] http://travel.ezinemark.com/medieval-towns-and-villages-in-spain-16bc87fb4dc.html

[26] "A project proposal" (http://almashriq.hiof.no/lebanon/400/410/412/elies_project/glimse_of_yesterday.html). Almashriq.hiof.no. . Retrieved 2010-03-28.

[27] "Village" (http://www.websters-online-dictionary.org/definition/english/vi/village.html#Definitions). Websters-online-dictionary.org. . Retrieved 2010-03-28.

[28] "Detailed map of Ohio" (http://www2.census.gov/geo/maps/general_ref/cousub_outline/cen2k_pgsz/oh_cosub.pdf) (PDF). United States Census Bureau. 2000. . Retrieved 2010-03-28.

[29] "Ohio Revised Code Section 703.01(A)" (http://codes.ohio.gov/orc/703.01). . Retrieved 2010-03-28.

[30] 2002 Census of Governments, Individual State Descriptions (http://www.census.gov/prod/2005pubs/gc021x2.pdf) (PDF)

External links

- Types of villages (anthropogenic biomes) (http://www.ecotope.org/anthromes/v1/guide/villages/)

First Partition of Poland

The **First Partition of Poland** or **First Partition of the Polish-Lithuanian Commonwealth** took place in 1772 as the first of three partitions that ended the existence of the Polish-Lithuanian Commonwealth by 1795. Growth in the Russian Empire's power, threatening the Kingdom of Prussia and the Habsburg Austrian Empire, was the primary motive behind this first partition. Frederick the Great engineered the partition to prevent Austria, jealous of Russian successes against Turkey, from going to war. The weakened Commonwealth's land, including that already controlled by Russia, was apportioned among its more powerful neighbors—Austria, Russia and Prussia—so as to restore the regional balance of power in Eastern Europe among those three countries. With Poland unable to effectively defend itself, and with foreign troops already inside the country, the Polish parliament (Sejm) ratified the partition in 1773 during the Partition Sejm convened by the three powers.

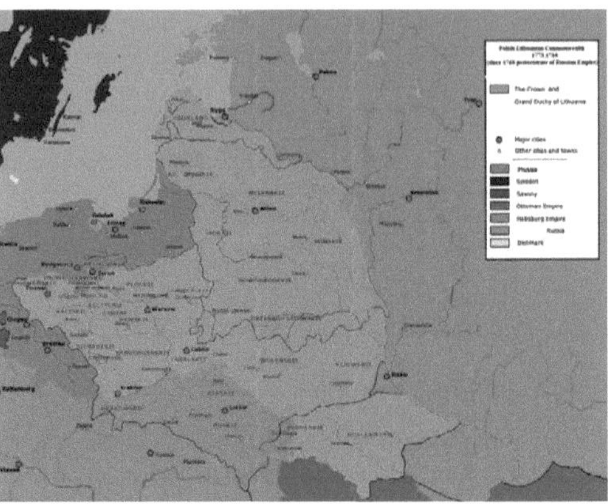

Polish-Lithuanian Commonwealth after the First Partition as a protectorate of Russian Empire 1773-1789

Picture of Europe for July 1772, satirical British plate

Background

In the late 17th century and early 18th century the Polish-Lithuanian Commonwealth had been reduced from the status of a major European power to that of a Russian protectorate (or vassal or satellite state), with the Russian tsar effectively choosing Polish-Lithuanian monarchs during the free elections and deciding the outcome of much of Poland's internal politics, for example during the Repnin Sejm, named after the Russian ambassador who unofficially presided over the proceedings.[1] [2]

The First Partition occurred after the balance of power in Europe shifted, with Russian victories against the Ottomans in the Russo-Turkish War (1768–1774) strengthening Russia and endangering Habsburg interests in that region (particularly in Moldavia and Wallachia). At that point Habsburg Austria started to consider waging a war against Russia.[3] [4]

France, friendly towards both Russia and Austria, suggested a series of territorial adjustments, in which Austria would be compensated by parts of Prussian Silesia, and Prussia in turn would receive Polish Ermland (Warmia) and parts of the Polish fief, Duchy of Courland and Semigallia—already under Baltic German hegemony. King Frederick II of Prussia had no intention of giving up Silesia gained recently in the Silesian Wars; he was, however, also interested in finding a peaceful solution—his alliance with Russia would draw him into a potential war with Austria, and the Seven Years' War had left Prussia's treasury and army weakened. He was also interested in protecting the weakening Ottoman Empire, which could be advantageously utilized in the event of a Prussian war either with Russia or Austria. Frederick's brother, Prince Henry, spent the winter of 1770–71 as a representative of the Prussian court at Saint Petersburg. As Austria had annexed 13 towns in the Hungarian Szepes region in 1769 (violating the Treaty of Lubowla),

The Troelfth Cake, a 1773 French allegory by Jean-Michel Moreau le Jeune for the First Partition of Poland.[a]

Catherine II of Russia and her advisor General Ivan Chernyshyov suggested to Henry that Prussia claim some Polish land, such as Ermland. After Henry informed him of the proposal, Frederick suggested a partition of the Polish borderlands by Austria, Prussia, and Russia, with the largest share going to the party most weakened by the recent changes in balance of power, Austria. Thus Frederick attempted to encourage Russia to direct its expansion towards weak and non-functional Poland instead of the Ottomans.[3] Austrian statesman, Wenzel Anton Graf Kaunitz, counter-proposed that Prussia take lands from Poland in return for relinquishing Silesia to Austria, but this plan was rejected by Frederick.

Although for a few decades (since the times of the Silent Sejm) Russia had seen weak Poland as its own protectorate,[1] Poland had also been devastated by a civil war in which the forces of the Bar Confederation attempted to disrupt Russian control over Poland.[3] The recent Koliyivschyna peasant and Cossack uprising in Ukraine also weakened the Polish position. Further, the Russian-supported Polish king, Stanisław August Poniatowski, was seen as both weak and too independent-minded; eventually the Russian court decided that the usefulness of Poland as a protectorate had diminished.[5] The three powers officially justified their actions as a compensation for dealing with troublesome neighbor and restoring order to Polish anarchy (the Bar Confederation provided a convenient excuse); in fact all three were interested in territorial gains.[6]

After Russia occupied the Danubian Principalities, Henry convinced Frederick and Archduchess Maria Theresa of Austria that the balance of power would be maintained by a tripartite division of the Polish-Lithuanian Commonwealth instead of Russia taking land from the Ottomans. Under pressure from Prussia, which for a long time wanted to annex the northern Polish province of Royal Prussia, the three powers agreed on the First Partition of Poland. This was in light of the possible Austrian-Ottoman alliance[7] with only token objections from Austria,[5] which would have instead preferred to receive more Ottoman territories in the Balkans, a region which for a long time had been coveted by the Habsburgs. The Russians also withdrew from Moldavia away from the Austrian border. An attempt of Bar Confederates to kidnap king Poniatowski on 3 November 1771 gave the three courts another pretext to showcase the "Polish anarchy" and the need for its neighbors to step in and "save" the country and its citizens.[8]

Partition begins

Already by 1769—71, both Austria and Prussia had taken over some border territories of the Commonwealth, with Austria taking Szepes County in 1769-1770 and Prussia incorporating Lauenburg and Bütow.[5] On February 19, 1772, the agreement of partition was signed in Vienna.[7] A previous agreement between Prussia and Russia had been made in Saint Petersburg on February 6, 1772.[7] Early in August Russian, Prussian and Austrian troops simultaneously entered the Commonwealth and occupied the provinces agreed upon among themselves. On August 5, the three parties signed the treaty on their respective territorial gains on the commonwealth's expense.[3]

The regiments of the Bar Confederation, whose executive board had been forced to leave Austria (which previously supported them[7]) after that country joined the Prusso-Russian alliance, did not lay down their arms. Many fortresses in their command held out as long as possible; Wawel Castle in Kraków fell only at the end of April;[7] [9] Tyniec fortress held until the end of July 1772;[10] Częstochowa, commanded by Kazimierz Pułaski, held until late August.[7] [11] In the end, the Bar Confederation was defeated, with its members either fleeing abroad or being deported to Siberia by the Russians.[12]

Division of territories

The partition treaty was ratified by its signatories on September 22, 1772.[7] It was a major success for Frederick II of Prussia:[7] [11] Prussia's share might have been the smallest, but it was also significantly developed and strategically important.[5] Prussia took most of Polish Royal Prussia, including Ermland, allowing Frederick to link East Prussia and Brandenburg. Prussia also annexed northern areas of Greater Poland along the Noteć River (the Netze District), and northern Kuyavia, but not the cities of Danzig (Gdańsk) and Thorn (Toruń).[3] The territories annexed by Prussia became a new province in 1773 called West Prussia. Overall, Prussia gained 36,000 km² and about 600,000 people. According to Jerzy Surdykowski Frederick the Great soon introduced German colonists on territories he conquered and engaged in Germanization of Polish territories[13] According to Christopher Clark 54 percent of the area and 75 percent of the urban populace were German-speaking Protestants.[14] In the next century this was used by nationalistic German historians to justify the partition,[14] but it was irrelevant to contemporary calculations; Frederick, dismissive of the German culture, was instead pursuing an imperialist policy, acting on the security interests of his state.[14] The new-gained territories connected Prussia with Germany proper, and were of major economic importance.[15] By seizing northwestern Poland, Prussia instantly cut off Poland from the sea,[15] and gained control over 80% of the Commonwealth's total foreign trade. Through levying enormous custom duties, Prussia accelerated the inevitable collapse of the Polish-Lithuanian state.[5]

Despite token criticism of the partition from Austrian Empress Maria Theresa,[5] [16] [17] Austrian statesman Wenzel Anton Graf Kaunitz considered the Austrian share an ample compensation; despite Austria being the least interested in the partition, it received the largest share of formerly Polish population, and second largest land share (83,000 km² and 2,650,000 people). To Austria fell Zator and Auschwitz (Oświęcim), part of Little Poland embracing parts of the counties of Kraków and Sandomierz (with the rich salt mines of Bochnia and Wieliczka), and the whole of Galicia, less the city of Kraków.[3]

Russia received the largest, but least-important area economically, in the northeast.[5] By this "diplomatic document" Russia came into possession of the commonwealth territories east of the line formed roughly by the Dvina, Drut, and Dnieper rivers — that section of Livonia which had still remained in commonwealth control, and of Belarus embracing the counties of Vitebsk, Polotsk and Mstislavl.[3] Russia gained 92,000 km² and 1,300,000 people, and reorganized its newly acquired lands into Yekaterinoslav Governorate and Mogilev Governorate.

By the first partition the Polish-Lithuanian Commonwealth lost about 211,000 km² (30% of its territory, amounting at that time to about 733,000 km²), with a population of over four to five million people (about a third of its population of 14 million before the partitions).[3] [18]

Aftermath

Rejtan - The Fall of Poland, oil on canvas by Jan Matejko, 1866, 282 x 487 cm, Royal Castle in Warsaw.

After having occupied their respective territories, the three partitioning powers demanded that King Stanisław August Poniatowski and the Sejm approve their action.[7] The king appealed to the nations of Western Europe for help and tarried with the convocation of the Sejm.[7] The European powers reacted to the partition with utmost indifference; only a few voices — like that of Edmund Burke — were raised in objection.[3] [7]

When no help was forthcoming and the armies of the combined nations occupied Warsaw to compel by force of arms the calling of the assembly, no alternative could be chosen save passive submission to their will. Those of the senators who advised against this step were threatened by the Russians, represented by the ambassador, Otto von Stackelberg, who declared that in the face of refusal the whole capital of Warsaw would be destroyed by them. Other threats included execution, confiscation of estates, and further increases of partitioned territory;[19] according to Edward Henry Lewinski Corwin, some senators were even arrested by the Russians and exiled to Siberia.[7]

The local land assemblies (Sejmiks) refused to elect deputies to the Sejm, and after great difficulties less than half of the regular number of representatives came to attend the session led by Marshals of the Sejm, Michał Hieronim Radziwiłł and Adam Poniński; the latter in particular was one of many Polish nobles bribed by the Russians and following their orders.[7] [20] This sejm became known as the Partition Sejm. In order to prevent the disruption of the Sejm via liberum veto and the defeat of the purpose of the invaders, Poniński undertook to turn the regular Sejm into a confederated sejm, where majority rule prevailed.[7] In spite of the efforts of individuals like Tadeusz Rejtan, Samuel Korsak, and Stanisław Bohuszewicz to prevent it, the deed was accomplished with the aid of Poniński, Radziwiłł, and the bishops Andrzej Młodziejowski, Ignacy Jakub Massalski, and Antoni Kazimierz Ostrowski (primate of Poland), who occupied high positions in the Senate of Poland.[7] The Sejm elected a committee of thirty to deal with the various matters presented.[7] On September 18, 1773, the committee formally signed the treaty of cession, renouncing all claims of the commonwealth to the lost territories.[7]

See also

* Administrative division of Polish territories after partitions
* Second Partition of Poland

Notes

a The picture shows the rulers of the three countries that participated in the partition tearing a map of Poland apart. The outer figures demanding their share are Catherine II of Russia and Frederick II of Prussia. The inner figure on the right is the Habsburg Emperor Joseph II, who appears ashamed of his action (although in reality he was more of an advocate of the partition, and it was his mother, Maria Theresa, who was critical of the partition). On his left is the beleaguered Polish king, Stanisław August Poniatowski, who is experiencing difficulty keeping his crown on his head. Above the scene the angel of peace trumpets the news that civilized eighteenth-century sovereigns have accomplished their mission while avoiding war. The drawing gained notoriety in contemporary Europe, with bans on its distribution in several European countries.

References

[1] Jerzy Lukowski, Hubert Zawadzki, A Concise History of Poland, *Cambridge University Press*, 2001, ISBN 0521559170, Google Print, p.84 (http://books.google.com/books?id=NpMxTvBuWHYC&pg=PA84&dq=Russia+Poland+protectorate& ei=ZUQWSObJOpDAygSs84W6Dw&sig=5f2ULjI3D3Ij2giHUFM5bobop7E)

[2] Hamish M. Scott, *The Emergence of the Eastern Powers, 1756-1775*, Cambridge University Press, 2001, ISBN 052179269X, Gooble Print, p.181-182 (http://books.google.com/books?ie=UTF-8&vid=ISBN052179269X&id=lc8EMD0JYUAC&pg=PA182&lpg=PA182& dq=Repnin+Poland&sig=VSKuu8NyPzm00Z6Rw1BNdGb0FP4)

[3] Poland, Partitions of. (2008). In Encyclopædia Britannica. Retrieved April 28, 2008, from Encyclopædia Britannica Online: http://www. britannica.com/eb/article-9060581

[4] Little, Richard. The Balance of Power in International Relations. Cambridge University Press, 2007. ISBN 9780521874885

[5] Poland. (2008). In Encyclopædia Britannica. Retrieved May 5, 2008, from Encyclopædia Britannica Online: http://www.britannica.com/ eb/article-28200 . Section: History > The Commonwealth > Reforms, agony, and partitions > The First Partition

[6] Sharon Korman, *The Right of Conquest: The Acquisition of Territory by Force in International Law and Practice*, Oxford University Press, 1996, ISBN 0198280076, Google Print, p.75 (http://books.google.com/books?id=ueDO1dJyjrUC&pg=PA75&dq=first+partition+of+ Poland+anarchy)

[7] Edward Henry Lewinski Corwin, *The Political History of Poland*, 1917, p. 310-315 (Google Print - public domain - full text online (http:// books.google.com/books?id=9foDAAAAYAAJ&dq))

[8] David Pickus (2001). *Dying with an enlightening fall: Poland in the eyes of German intellectuals, 1764-1800* (http://books.google.com/ books?id=iZYTfgZCb7cC&pg=PA35). Lexington Books. p. 35. ISBN 978-0-7391-0153-7. . Retrieved 4 December 2011.

[9] (**Polish**) Halina Nehring Kartki z kalendarza: kwiecień (http://www.opcja.pop.pl/numer28/28kal.html)

[10] (**Polish**) Tyniec jako twierdza Konfederatów Barskich (http://www.naszradziszow.com/pl/art45.htm)

[11] Norman Davies, *God's Playground: A History of Poland in Two Volumes*, Oxford University Press, 2005, ISBN 0199253390, Google Print, p.392 (http://books.google.com/books?id=b912JnKpYTkC&pg=RA1-PA392&vq=Frederick+partition&dq=Frederick+partition+ Davies&lr=&source=gbs_search_s&sig=jzdAqAdKtMUuft_TWEFglh-ktRY)

[12] Norman Davies, *Europe: A History*, Oxford University Press, 1996, ISBN 0198201710, Google Print, p.664 (http://books.google.com/ books?id=jrVW9W9eiYMC&pg=PA664&dq=Polish+Bar+Siberia&lr=&as_brr=3&sig=-Qnx4AN3txLWtn9WTfaMLO0RM4w)

[13] Duch Rzeczypospolitej Jerzy Surdykowski - 2001 Wydawn. Nauk. PWN, 2001, page 153

[14] Christopher M. Clark (2006). *Iron kingdom: the rise and downfall of Prussia, 1600-1947* (http://books.google.com/ books?id=4LPODzLgDVEC&pg=PA233). Harvard University Press. pp. 233–. ISBN 9780674023857. . Retrieved 17 February 2011.

[15] Christopher M. Clark (2006). *Iron kingdom: the rise and downfall of Prussia, 1600-1947* (http://books.google.com/ books?id=4LPODzLgDVEC&pg=PA232). Harvard University Press. pp. 232–. ISBN 9780674023857. . Retrieved 17 February 2011.

[16] Frederick II of Prussia wrote about the participation of Maria Theresa in the first partition in a letter: "The Empress Catherine and I are simple robbers. I just would like to know how the empress calmed down her father confessor? She cried, when she took; the more she cried, the more she took!?" Davies, p.390 (http://books.google.com/books?id=b912JnKpYTkC&pg=RA1-PA390&vq=Frederick+partition& dq=Frederick+partition+Davies&lr=&source=gbs_search_s&sig=O6URburdADQO6xdxrMT76MbRxms)

[17] Sharon Korman, *The right of conquest: the acquisition of territory by force in international law and practice*, Oxford University Press, 1996, ISBN 0198280076, Google Print, p.74 (http://books.google.com/books?id=ueDO1dJyjrUC&pg=PA74&dq=Frederick+II+ partition+1772&lr=&as_brr=3&sig=DNoG7x4tMGDWRuT4Uihr-SDB7Z4#PPA74,M1)

[18] Jerzy Lukowski, Hubert Zawadzki, A Concise History of Poland, *Cambridge University Press*, 2001, ISBN 0521559170, Google Print, p.97 (http://books.google.com/books?id=NpMxTvBuWHYC&pg=PA96&vq=demographic+estimates&dq=Russia+Poland+protectorate&

source=gbs_search_s)

[19] Historia Encyklopedia Szkolna Wydawnictwa Szkolne i Pedagogiczne Warszawa 1993 page 525
    "Opponents were threatened with executions, increase of partitioned territories, and destruction of the capital"

[20] Jerzy Jan Lerski, Piotr Wróbel, Richard J. Kozicki, *Historical Dictionary of Poland, 966-1945*, Greenwood Publishing Group, 1996, ISBN
    0313260079, Google Print, p.466 (http://books.google.com/books?id=S6aUBuWPqywC&pg=PA466&dq=Adam+PoniÅski+bribed&
    sig=UMTRdzAO_OSc9ytZnQ-_WIOVXZM)

Further reading

- Herbert H. Kaplan, *The First Partition of Poland*, Ams Pr Inc (June 1972), ISBN 0404036368
- Tadeusz Cegielski, Łukasz Kądziela, *Rozbiory Polski 1772-1793-1795*, Warszawa 1990
- Władysław Konopczyński *Dzieje Polski nowożytnej*, t. 2, Warszawa 1986
- Tomasz Paluszyński, *Czy Rosja uczestniczyła w pierwszym rozbiorze Polski czyli co zaborcy zabrali Polsce w trzech rozbiorach. Nowe określenie obszarów rozbiorowych Polski w kontekście analizy przynależności i tożsamości państwowej Księstw Inflanckiego i Kurlandzkiego, prawnopaństwowego stosunku Polski i Litwy oraz podmiotowości Rzeczypospolitej*, Poznań 2006.
- S. Salmonowicz, *Fryderyk Wielki*, Wrocław 2006
- Maria Wawrykowa, *Dzieje Niemiec 1648-1789*, Warszawa 1976
- Editor Samuel Fiszman, *Constitution and Reform in Eighteenth-Century Poland*, Indiana University Press 1997 ISBN 0-253-33317-2
- Jerzy Lukowski *Liberty's Folly The Polish-Lithuanian Commonwealth in the Eighteenth Century*, Routledge 1991 ISBN 0-415-03228-8
- Adam Zamoyski *The Last King of Poland*, Jonathan Cape 1992 ISBN 0-224-03548-7

External links

- James Fletcher First Partition Of Poland (http://history-world.org/poland.htm)
- D. B. Horn, review of The First Partition of Poland by Herbert H. Kaplan, The English Historical Review, Vol. 79, No. 313 (Oct., 1964), pp. 863-864 (review consists of 2 pages), JSTOR (http://www.jstor.org/pss/560615)
- O. Halecki, Reviewed work(s): British Public Opinion and the First Partition of Poland. by D. B. Horn, American Slavic and East European Review, Vol. 4, No. 3/4 (Dec., 1945), pp. 205-207 JSTOR (http://www.jstor.org/pss/2491775)
- J. T. Lukowski, *Guarantee or Annexation: a Note on Russian Plans to acquire Polish Territory prior to the First Partition of Poland*, Historical Research, Vol. 56 Issue 133 Page 60 May 1983, (http://www.blackwell-synergy.com/doi/pdf/10.1111/j.1468-2281.1983.tb01759.x?cookieSet=1)
- The Three Partitions, 1764-95: First Partition (http://historymedren.about.com/library/text/bltxtpoland13.htm), Library of Congress Country Study
- The Period of Partitions (1772-1918) (http://info-poland.buffalo.edu/web/history/partitions/link.shtml) - resources
- (Polish) Photos of some contemporary documents (http://dziedzictwo.polska.pl/katalog/skarb,Traktaty_polsko-austriackie_dotyczace_I_rozbioru_zawarte_dnia_18_IX_1773_roku_i_16_III_1775_roku,gid,111073,cid htm)
- (Polish) Polish-Russian (http://pl.wikisource.org/wiki/Traktat_pierwszego_podziaÅu_pomiÄdzy_RzeczÄ pospolitÄ _i_RosjÄ _(1773)) and Polish-Austrian (http://pl.wikisource.org/wiki/Traktaty_polsko-austriackie_dotyczÄ ce_I_rozbioru) treaties of the First Partition
- Map of the 1st Partition (http://unimaps.com/poland-1stPart/index.html)

Article Sources and Contributors

Rokitnik, Lidzbark County *Source*: http://en.wikipedia.org/w/index.php?title=Rokitnik%2C_Lidzbark_County *Contributors*: 1 anonymous edits

East Prussia *Source*: http://en.wikipedia.org/w/index.php?title=East_Prussia *Contributors*: 21655, 52 Pickup, AAA!, ABE, Acategory, Adam Bishop, AdjustShift, Ahoerstemeier, AlexiusHoratius, Altenmann, Andries, Anthony Appleyard, Arjun01, Ataylor, Auntof6, Balcer, Before My Ken, Beland, Bkell, Bledmeister, Branddobbe, Burschenschafter, C. A. Russell, Caius2ga, Cautious, Cemsentin1, CharonX, Chaumot, Chowbok, Chris 73, Chris the speller, Christchurch, Colonel Mustard, Colonies Chris, CommonsDelinker, Conscious, Cyrius, D6, DaQuirin, Darkildor, Darklilac, David Liuzzo, DeirYassin, Dellijks, Deqon, Der Eberswalder, DerHexer, Dewritech, DocWatson42, Domino theory, DrFrench, Droll, Dtremenak, Dzey, Ekotkie, Eon, Europa, Ev, FMB, Fitzwilliam, Formeruser-83, Fuhghettaboutit, Funnyhat, Gabbe, Gaius Cornelius, Gauss, Ghepeu, Ghewgill, Grocer, Ground Zero, Guinness man, Gunternitsch, Hallmark, Harald82, Hayden120, Hectorian, Hellenica, HerkusMonte, Heron, Hmains, Huaiwei, Ineuw, Ingolfson, Iohannes Animosus, Irredenta, Iulius, Jaan, Jadger, Jj137, John Quiggin, Jor, Juditten, Ksenon, Kwamikagami, Kyle sb, LA2, LOL, Laurence Portacio, Legaleagle86, Lightmouse, Lysy, Maiatues, Mateo LeFou, Matthead, MatttK, Maximus Rex, Mervyn, Mickey195, Mild Bill Hiccup, Mnmazur, Molinari, Molobo, Morton devonshire, Mtaylor848, MyMoloboaccount, N-true, NachOking, Neg, Netoholic, Nico, Nixer, Olessi, OwenBlacker, Ozzieboy, Palica, Patrick, Paul Benjamin Austin, PeterBln, Petri Krohn, Philip Baird Shearer, PigFlu Oink, Piotrus, Postdlf, R'n'B, Rachsitten, Ran, Rich Farmbrough, RickK, Robertgreer, RottweilerCS, Ruhrjung, SM, Sannse, Sardanaphalus, Sca, Schwartz und Weiss, Schwyz, Sciurinæ, Sdornan, Seeaxid, Shauri, Skäpperöd, Slashme, Smith2006, Something12356789, Soulpatch, Space Cadet, Stronach, Stubblyhead, The Rationalist, ThePiedCow, Timc, Timeineurope, Tirid Tirid, Tobias Conradi, Toby Bartels, UW, Valip, Volunteer Marek, Waltpohl, Wareq, Weisbrod, Welkinridge, Wighson, Witkacy, Wmahan, Writtenright, YUL89YYZ, Yelizandpaul, Yms, Yopie, Zack Holly Venturi, Zestauferov, Île flottante, 240 anonymous edits

Crown of the Kingdom of Poland *Source*: http://en.wikipedia.org/w/index.php?title=Crown_of_the_Kingdom_of_Poland *Contributors*: Ajh1492, Bonás, Cautious, CommonsDelinker, EdGl, Elonka, Greg park avenue, Grutness, Halibutt, Hmains, Iaroslavvs, Irpen, JamesAM, Kpalion, Ksbrown, Mandarax, Mathiasrex, Mnplastic, Nihil novi, Omnipaedista, Only, PANONIAN, Pearle, Piotrus, Rjwilmsi, Robofish, Romuald Wróblewski, TexasAndroid, Ubudoda, Valentinian, Waterloo1974, Woohookitty, 38 anonymous edits

Kiwity, Warmian-Masurian Voivodeship *Source*: http://en.wikipedia.org/w/index.php?title=Kiwity%2C_Warmian-Masurian_Voivodeship *Contributors*: 1 anonymous edits

Olsztyn *Source*: http://en.wikipedia.org/w/index.php?title=Olsztyn *Contributors*: AN(Ger), Abtinb, Ak47K, Andreanrc, Andy4226uk, Astronaut, Balcer, Barticus88, Bassbonerocks, Beobach972, Berek, BlackJack, Bobblewik, Boris51, Borya01, CORNELIUSSEON, Caius2ga, Cautious, Cel 84, Cezary Okupski, Chanheigeorge, Chi45cken, Chris G, CommonsDelinker, Curps, Danny, Darius Dhlomo, Darwinek, David Lauder, Derek Ross, Dimadick, Discospinster, Dzey, EAman, EBL, Ecphora, Elonka, Emax, Enchanter, Fingers-of-Pyrex, Gargolla, Geniu, GorgeCustersSabre, Green Giant, Gzornenplatz, Halibutt, HerkusMonte, Hiuppo, Hmains, Jacurek, Jim, Josh a brewer, Just plain Bill, Karol Langner, Kelisi, Keyboard warrior killer, Kim Traynor, Kitauga, Kotniski, Kpalion, Kpjas, Ksenon, Leoboudv, Lightmouse, Lysy, Maasteerpl, Marek69, Matthead, Meursault2004, Michael Romanov, Mirek46, Molobo, MyMoloboaccount, Nickkid5, Nihil novi, Nihiltres, Nina Subocz, Ohconfucius, Olessi, Parslad, Paxsimius, PeterBln, Piotr Mikołajski, Poeticbent, PolishPoliticians, R'n'B, R9tgokunks, Rymano, Sarcelles, Sardanaphalus, Sca, Schwartz und Weiss, Semper malus, Shoeofdeath, Sicherlich, Smith2006, Space Cadet, Stan Shebs, Stanislaw Albinowski, StoneProphet, SwisterTwister, Szopen, Tabletop, Tamfang, Tdls, Ten-pint, The PIPE, ThomasPusch, Tirid Tirid, Toby Bartels, Tsca, Ulf Heinsohn, Umix, Volunteer Marek, Widefox, WojPob, Woohookitty, YUL89YYZ, You forgot Poland, 191 anonymous edits

Lidzbark Warmiński *Source*: http://en.wikipedia.org/w/index.php?title=Lidzbark_Warmi%C5%84ski *Contributors*: Adam78, Angr, Balcer, Bonás, Cautious, Chris G, CommonsDelinker, Curps, Darwinek, Dimadick, Elonka, Emax, Everyking, Fawcett5, Finell, Goethicus, Graham87, Gzornenplatz, Halibutt, Hiuppo, James Russiello, Jor, Kotniski, Kwamikagami, LilHelpa, Lysy, Marek69, Margoz, Matthead, Molobo, Olessi, Piotrus, PolishPoliticians, ProudPomeranian, R9tgokunks, Schwartz und Weiss, Sicherlich, Smith2006, Space Cadet, Steve64, TOR, Volunteer Marek, Wik, Witkacy, 48 anonymous edits

Lidzbark Warmiński *Source*: http://en.wikipedia.org/w/index.php?title=Lidzbark_Warmi%C5%84ski *Contributors*: Adam78, Angr, Balcer, Bonás, Cautious, Chris G, CommonsDelinker, Curps, Darwinek, Dimadick, Elonka, Emax, Everyking, Fawcett5, Finell, Goethicus, Graham87, Gzornenplatz, Halibutt, Hiuppo, James Russiello, Jor, Kotniski, Kwamikagami, LilHelpa, Lysy, Marek69, Margoz, Matthead, Molobo, Olessi, Piotrus, PolishPoliticians, ProudPomeranian, R9tgokunks, Schwartz und Weiss, Sicherlich, Smith2006, Space Cadet, Steve64, TOR, Volunteer Marek, Wik, Witkacy, 48 anonymous edits

Lidzbark County *Source*: http://en.wikipedia.org/w/index.php?title=Lidzbark_County *Contributors*: Balcer

Gmina Kiwity *Source*: http://en.wikipedia.org/w/index.php?title=Gmina_Kiwity *Contributors*:

Village *Source*: http://en.wikipedia.org/w/index.php?title=Village *Contributors*: 1223Sallybride, 21655, A.Hassen, AKeen, Aarohan Mettu, Abductive, AdamDeanHall, Adamhauner, Airjatt, Aitias, Akrabbim, Alaexis, Ale jrb, AlefZet, Aleksandr Grigoryev, Alensha, Altenmann, Amberrock, Aminullah, Andrewpmk, Andycjp, Animum, Apalamarchuk, Apokrif, Arpingstone, Arthena, AussieLegend, Azreey, Bejinhan, Beland, Belirac, Berig, Big Bird, Bill37212, Biruitorul, Bjh21, Bkonrad, Bldonne, Bobrayner, BocoROTH, Bogdan, Bogdangiusca, Bomac, Bossmanham13, Bossmanham98, Branka France, BreakingFree40, Brian the Editor, Brian0918, Buistr, Butko, C.Fred, Caiaffa, CalJW, CalumMcA, Canley, Capricorn42, Cassowary, Catgut, Catslisten, Cecropia, Cenarium, Ceriy, Cgoodwin, Chasnor15, Chentchen, Chirag, Chris j wood, Christopher.e.dunn, Chzz, Ciudad jardin, Cntras, Colonies Chris, Conleyb15, Conorbrady.ie, Corregere, Cosmic Latte, CrimsonBlue, Cwolfsheep, D6, DITWIN GRIM, Da-drumer, DailyWikiHelp, Daniel Case, Daniel Quinlan, David Biddulph, David Kernow, Dbg1233, Ddstretch, DePiep, Deathawk, Deeptrivia, Den fjättrade ankan, Deqon, Deror avi, Desiphral, Dezaree, Dieter Simon, Discospinster, Doctor Whom, Donarreiskoffer, Doseiai2, Dottod, Dpkipling, Dysepsion, E ponti, E0steven, Ehrenkater, Elassint, Eleassar, Ellywa, Elockid, Elroygoh02, Emyr93, Epipelagic, Erianna, Eu.stefan, EugeneZelenko, EurekaLott, Extra999, Ezhiki, Fang 23, Farosdaughter, Fat pig73, Flewis, Francis Tyers, Francs2000, Fred Bauder, Frosted14, Fujurcitook, Funandtrvl, FunkyFly, Futurebird, Gardar Rurak, Garhowell, Gegnome, Gfoley4, Gingermint, Ginkgo100, Giro720, Glenn, Globalphilosophy, GoOdCoNtEnT, Gogo Dodo, Gorai, Grafikm fr, GrahamSmithe, Gtswiki, Guanaco, Guss2, Gustavo Szwedowski de Korwin, Gz33, Hahnchen, Halsteadk, Harisu1988, Hayabusa future, Hebrides, Helenmwarwick, HenryLi, HighSpeed-X, Hmains, Hroðulf, Hu12, Hydrogen Iodide, IGeMiNix, IShadowed, IW.HG, Ianisbored, Indiaforever, Ineffable3000, Inwind, Isam, IstvanWolf, IvanKlinko, J.delanoy, JHunterJ, JPSheridan, Jahangard, James086, Janeleopold, Jared Preston, Jcuk, JdeJ, Jeremy Bolwell, Jerrch, Jno, Jongleur100, Julesd, Julia W, JuneGloom07, KR3ColinB, Kajasudhakarababu, Kamiekam, Keelanrush, Kelisi, Kelovy, Khalid Mahmood, Khanvez, Khazar2, Kikos, Kimberry352, Kironbd07, KirrVlad, Kjramesh, Kman543210, Kmsasidhar, Komalrajiana, Ktr101, Kurruption, Kwamikagami, Larucio, Laurinavicius, Ldonna, Leonig Mig, Levineps, Lew121, LiDaobing, Lightmouse, LilHelpa, Linaduliban, Little muddy funkster, Littlealien182, Llywrch, Lonewolf BC, Lord Geznikor, Lozleader, Luk, Lulu1127, Lumos3, Lynda Finn, MacTire02, Maglame, Maitch, Majorly, Malick78, Mallakottai, Mani1, Manly 2008, Marek69, Mariano12 1989, Mariwan zanko, Mark91, Markmark12, Materialscientist, Mayooranathan, Mdzafri, Meldor, Mfo7787, MickMacNee, MisfitToys, MisterVodka, Modulatum, Moebiusuibeom-en, Moonriddengirl, Mrmuk, Ms. Hayden and Ms. S, Mtaylor848, Mvjs, Mwalcoff, Mxn, NE2, Naturenet, Naveen Sankar, NerdyScienceDude, Nethergreen55, NewEnglandYankee, Nickshanks, Nistra, Noclevername, Nricardo, Nyttend, Olive, Old Moonraker, Orangemike, OscarKosy, P.Marlow, Parkwells, Patrick, PaulBannister, Pawyilee, Penrithguy, Peter E. James, PeterCScott, Pgan002, Phagopsych, Pharaoh of the Wizards, PhilKnight, Pinethicket, PleaseStand, Pokrajac, Ppeace, Pug1776, Puneet 348, Quaysys, R'n'B, RDF, Raeky, Randhirreddy, RandomAct, Rcingham, Realteacher, Reeses14, Reinyday, Rich Farmbrough, Rickyrab, Risker, Rjm at sleepers, Robdav69, Rocket71048576, Rwxrwxrwx, Saccerzd, Sadrettin, Saga City, Same7same7, SamuraiClinton, Samyo, Sansonic, Sardanaphalus, Schzmo, ScottSteiner, Seb az86556, Sem10efg, Sephiroth BCR, Sfu, Shakingjoker117, Shantavira, Sheddybey, Shell Kinney, Shoeofdeath, SilkTork, Sjakkalle, Sketchmoose, Sljaxon, Sole Soul, Speciate, StephenDawson, Steven Zhang, Stevo D, StewartNetAddict, Stifle, Student BSMU, Suffusion of Yellow, Suprgye, Surya Prakash.S.A., Szambo, THATSBETTER, Taamu, Tabletop, Tbhotch, Teinesavaii, Tekxtinct, TenIslands, Tenth Plague, The Thing That Should Not Be, The Transhumanist, Theranos, Thomasrflagel, Tide rolls, Tjanke, Tobby72, Tobias Conradi, Tpbradbury, Triona, Tsuchiya Hikaru, TurkChan, Twanner921, Twin Bird, Tytyty456, Ukabia, UnicornTapestry, Untifler, Utcursch, V2k, VMS Mosaic, Vegaswikian, Vhdwikiuser, Vikramdidawat, Vmenkov, WOSlinker, Walex03, Warofdreams, Wayland, Wayne Slam, Wendy1991, West Brom 4ever, Whytecypress, Wiki Wikardo, Wikijens, William Allen Simpson, William Avery, Winhunter, Wknight94, WojPob, Woohookitty, WorldWide Update, Wrelwser43, Xiaoyu of Yuxi, Xyz or die, Yakudza, Yonkie, Yupik, Zacharie Grossen, Zhou Yu, Zoeyzazu, Zollerriia, Zotel, Александър, პაატა ა, , , , 444 anonymous edits

First Partition of Poland *Source*: http://en.wikipedia.org/w/index.php?title=First_Partition_of_Poland *Contributors*: Anthony Appleyard, Ardfern, Bejnar, Biruitorul, Bwilkins, CN3777, Czalex, Discospinster, Estlandia, Eumolpo, France3470, Good Olfactory, Grahamec, Green Giant, HerkusMonte, Hires an editor, Irpen, Jd2718, Jniech, Koavf, Lajbi, Lokyz, M.K, Mathiasrex, Matthead, Michael Hardy, Molobo, MyMoloboaccount, NawlinWiki, Olessi, Pernambuko, Piotrus, Pleckaitis, Renata3, Russ3Z, Russavia, Skip237, Space Cadet, Top.Squark, Vecrumba, Woohookitty, პაატა ა, 23 anonymous edits

Image Sources, Licenses and Contributors

Printed by Books on Demand GmbH, Norderstedt / Germany